BRIDGES

To Ann and my mother

BRIDGES
Three Thousand Years of Defying Nature
David J. Brown

First published in Great Britain by Mitchell Beazley
an imprint of Octopus Publishing Group Ltd
2-4 Heron Quays
London E14 4JP

Executive Editor Frances Gertler
Art Director Tim Foster
Art Editors Rozelle Bentheim, John Grain
Associate Editor James Chambers
Assistant Editors Alison Macfarlane, Katie Piper
Picture Research Judy Todd
Illustrations (colour) Simon Miller
Illustrations (black and white) Coral Mula
Index Hilary Bird
Production Sarah Schuman, Michelle Thomas

A CIP catalogue record for this book is available from
the British Library
ISBN 1 8400 138 0

Produced by Toppan Printing Co. Ltd
Printed in China
Typeset in Frutiger by SX Composing Ltd

Cover Pictures
Front: The Forth Rail Bridge, Scotland (Richard
McConnell)
Back: The first Westminster Bridge (drawn by Simon
Miller)
Picture opposite: Suspension bridge under
construction near Rovrik, in north Frandelag,
mid-Norway (ZEFA/Rossenbach)

BRIDGES

THREE THOUSAND YEARS OF DEFYING NATURE

DAVID J. BROWN

MITCHELL BEAZLEY

contents

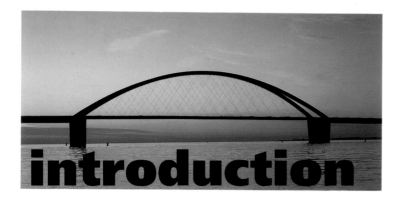

introduction

You will probably walk, drive, or be driven across at least one bridge today. So what? Isn't that all they are there for? Just to get "A" (you, other people, cars, trucks, trains, even boats and planes) from "B" across "C" to "D" as quickly and as easily as possible? Unlike a house, office, concert-hall or hospital, bridges don't have to accommodate complex, developing, long-term human activities, needs and services; people don't live, work, get entertained or cured in bridges. The better a bridge is at doing its job, the more transitory, literally, is your experience of it.

But is that all there is to the subject? If it is, this book wouldn't need to have been written. Although the initial question to be asked about any bridge is "Why?" (to which the answers will be provided by anthropologists, historians, sociologists or economists), as soon as the need for a bridge is apparent the next question becomes "How?", and that is what this book is all about.

The way builders and engineers have confronted and dealt with this question forms one of the most fascinating technological sagas in history. Although in the 20th century we have come increasingly to fear and distrust technology and its consequences – despite our ever-growing reliance on it – I hope this book shows at least one area of "big engineering" to be interesting and also benign; not for nothing has the phrase "building bridges" become a metaphor for all kinds of positive human activity – co-operation rather than conflict; helping, not hindering; linking, not sundering.

Bridges demonstrate that the human race has always used technology. A Neanderthal cutting down a log to cross a stream was using technology, however primitive; so were the medieval builders toiling for decade after decade to place the crude spans of Old London Bridge across the fast-flowing Thames; and the 19th-century engineer Sir John Rennie, using an early steam engine to construct the foundations for the replacement New London Bridge; as well as the modern builders who supplanted that bridge with smooth prestressed-concrete spans only 20 years ago. From prehistoric times to the present, the line of technology is unbroken, but what has changed is its level of sophistication and comprehensibility.

If you have driven across, say, San Francisco's Golden Gate or England's Humber Bridge, and wondered how it was built or stays up, this book might help you understand a little more about the way such fantastic structures work. But technical ingenuity is not the whole story. I hope that once you start looking *at* bridges rather than just crossing them, you will be continually struck (as I have been all over again while writing this book) by how beautiful so many of them can be – and by the range this beauty can take. These assemblies of rope, wood, stone, iron, steel, concrete and even plastics may not merely cross an industrial or inner city wasteland, or skirt around a beautiful landscape as unobtrusively as possible – they can also crown great natural splendour. It is virtually impossible now to visualize Sydney Harbour in Australia, the Firth of Forth in Scotland, or the Salgina Gorge in Switzerland, without the great bridges that span them, and it surely can't be said that they have defiled nature.

So follows the final point – great bridges demonstrate that there need be no conflict between technology and "art", but rather a fusion between the two. I hope this book also goes some way to show how great designers, from the Romans to the present day have combined technical mastery, consciously or unconsciously, with an artistic sensibility, to create not just structures to be crossed as quickly as possible, but masterpieces by any standard. The greatness of the bridge designer lies in the interpenetration of his or her technical mastery and creativity, with the one working with, for, and necessary to the other, to achieve the best-possible whole.

So if this book makes readers experience great bridges for the first time, it will have been well worth while. And if it makes some look twice and ponder as they pass under a slender arch – or an ugly slab – on a motorway or freeway, and wonder whether it had to look like that and if so, why?, and if not, why a better design job wasn't made of it, then it will have served a further purpose.

Writing a book such as this is like going on a journey of discovery, but one where the traveller cannot linger very long in any single place, so I have credited the invaluable records of more leisurely, detailed and specific "travellers" listed in the bibliography. Personal thanks go to Barbara Tait and Annette O'Brien (successive librarians), and others, Simon White, Julian Dawson, Heide Pirwitz, Marit Tronslin and Denis O'Kelly of Ove Arup Partnership Library for pushing many useful nuggets of information from current periodicals under my nose, and to Lisa Harris and the resources of that Library generally; to Mike Chrimes and his equally resourceful Library of the Institution of Civil Engineers; to Povl Ahm, Sir Jack Zunz, John Martin and many others at Ove Arup Partnership for their interest and encouragement; and also to Ken and the others for livening up dull, wet, wintry Sundays.

Much of the final chapters could not have been written and illustrated without the help of: Hamish Douglas, Dyckerhoff & Widmann; Chris Dixon, Dartford River Crossing; Hans Thomas Øderud, Norwegian Directorate of Public Roads Bridge Department; Thomas G.Lovett, Greiner Inc., Tampa, Fla; Bruce Burnett; Dorothy Kerr, BC Transit; Christian Menn; Frank Rowley, Tony Gee & Partners; Brian Richmond, G.Maunsell & Partners; Pat Murtagh, Acer Consultants; Anthony Tischauer, Calatrava Valls SA; Paolo Rosselli; Kensaku Hata and Masao Hisato, Honshu-Shikoku Bridge Authority; Anne Marie Richards, Royal Danish Embassy, London; Lise Uldal, COWIconsult; Jacob Vestergaard, Storebælt; Gabriele Del Mese; Mirella Refghi, Stretto di Messina SpA; and Dr.T.Y. Lin, T.Y. Lin International.

Finally, many thanks to Michael Bussell, Angus Low and Bill Smyth, who read portions of the text and made numerous valuable comments and suggestions; also, at Mitchell Beazley, to John Grain, Judy Todd, Rozelle Bentheim and in particular Frances Gertler, for her constant enthusiasm, encouragement and patience; and to Ann, without whom... she knows.

David J.Brown

CHRONOLOGY

(Dates are of completion, except where otherwise stated)

BC

Pre-2000BC — Arch construction techniques used in Mesopotamia, Egypt and elsewhere, but not in bridge-building. Primitive suspension bridge techniques probably known and used in China, India and elsewhere. Probable earliest use of floating pontoon and cantilever bridges in China.

15thC BC — Corbelled arches used in Mycenean "beehive" tombs.

7thC BC — Corbelled stone aqueduct, at Jerwan, Mesopotamia.

7-6thC BC — Bridge built across the Euphrates at Babylon.

6thC BC — Floating pontoon bridges built across the Danube and Bosphorus by Mandrocles of Samos.

480BC — Xerxes builds a pontoon bridge of over 600 ships across the Hellespont.

3rdC BC — Earliest records of Chinese bamboo and iron chain suspension bridges.

2ndC BC — Romans build masonry bridges over foundations secured in cofferdams.

c.62BC — Ponte Quattro Capi (Pons Fabricus) built in Rome.

c.55BC — Julius Caesar erects timber trestle bridge across the Rhine.

1stC BC — Vitruvius writes *De architectura*.

AD

before AD14 — Construction of Pont du Gard, France, and Pons Augustus, Rimini.

c.AD100 — Trajan builds aqueduct at Segovia, and bridges at Alcántara, Spain and across the Danube.

c.AD134 — Pons Aelius or Ponte Sant'Angelo built in Rome.

AD605 — Segmental-arched An Ji Bridge built in China.

10th-12thC — Heyday of Romanesque architecture, based on the semi-circular arch.

mid-12thC — Gothic style, based on the pointed arch, begins to replace Romanesque.

1188 — 20/21-arch Pont d'Avignon.

1210 — Old London Bridge: 19 pointed arches.

early 13thC — Villard de Honnecourt designs timber truss bridge.

1218 — Chain-suspended Twärenbrücke, Switzerland.

1345 — Segmental-arch Ponte Vecchio begun in Florence.

1357 — Charles Bridge begun across Danube in Prague.

1371 — The longest masonry arch span of the Middle Ages, 72m (237ft), built at Trezzo by the Duke of Milan.

1507 — Pont Notre-Dame, first stone bridge in Paris.

1511 — Birth of Bartolommeo Ammanati.

c.1512 — Birth of Antonio da Ponte.

1518 — Birth of Andrea Palladio.

1567 — Ponte Santa Trinità, Florence, built by Ammanati.

1570 — Palladio's *Four Books of Architecture* illustrates types of timber truss bridges.

1591 — Rialto Bridge, Venice, built by da Ponte.

1595-1617 — Verantius' bridge designs in *Machinae Novae*.

c.1600-40 — Allahverdi Khan and Khaju Bridges built in Isfahan.

1708 — Birth of Jean-Rodolphe Perronet.

1750 — Westminster Bridge: London's first crossing since Old London Bridge.

1755-60 — Timber arch bridges spanning over 61m (200ft) built at Schaffhausen, Reichenau and Wettingen by the Grubenmann brothers.

1757 — Birth of Thomas Telford.

1761 — Birth of John Rennie.

1772 — Perronet builds the Pont de Neuilly across the Seine.

1776 — World's longest suspended span of 200m (660ft), in bamboo, built in Szechuan province, China.

1779 — Ironbridge at Coalbrookdale: the first cast-iron arch.

1791 — Perronet builds Pont de la Concorde in Paris.

1796 — Iron arch Wear Bridge, Sunderland, Buildwas Bridge, and iron-trough Longdon-on-Tern Aqueduct.

1800 — First rigid suspension bridge built in USA by Judge James Finley.

1803 — Birth of Robert Stephenson.

1805 — Pontcysyllte Aqueduct built by Telford.
Timothy Palmer erects timber truss, covered "Permanent Bridge" over Schuylkill River, Pennsylvania.

1806 — Birth of John Roebling and Isambard Kingdom Brunel.

1808 — Finley patents suspension system.

1810 — Birth of Charles Ellet, designer of Wheeling Bridge, Ohio.

1811 — Lewis Wernwag's timber arch/truss "Colossus" Bridge built over the Schuylkill.

1815 — Iron arch Craigellachie and Bonar Bridges built by Telford in Scotland.

1815 — First use of Burr-type truss in USA.

1817 — First Waterloo Bridge, in masonry, built by Rennie.

1819 — First Southwark Bridge, in cast iron, built by Rennie.

1820 — Capt. Samuel Brown builds 137m (449ft) span suspension bridge across the Tweed.
Patent of Town timber lattice truss system in USA.
Birth of James Eads, designer of Mississippi Bridge.

1822-40 — Several hundred wrought-iron wire suspension bridges built in Europe by the Seguins, Dufour, and others.

1824-31 — Old London Bridge demolished and New London Bridge built, in masonry, to Rennie's design.

1826 — Telford builds the iron-chain Menai Straits suspension bridge, and Conway Bridge.

1830 — Patent of Long timber truss system in USA.

1831 — Vicat first proposes aerial spinning of wire cables.

1832 — Birth of Gustave Eiffel.

1834 — "Grand Pont Suspendu", the then world's longest suspension bridge built by Chaley at Fribourg in Switzerland.

1840 Patent of Howe cast-iron/timber truss system, USA.

1843 Birth of François Hennebique, pioneer of reinforced concrete.

1849 High Level Bridge, Newcastle, designed by Stephenson.
Ellet's record-breaking Wheeling suspension bridge over the Ohio – the first "thousand-footer".

1850 Robert Stephenson builds tubular-iron Conway and Britannia Bridges.
Collapse of Bass-Chaîne suspension bridge in France.

1855 John Roebling's double-deck road/rail suspension bridge at Niagara.

1859 Opening of tubular suspension Royal Albert Bridge, Saltash, built by Brunel.

1866 Roebling builds world-record suspension span, 322m (1,057ft), over the Ohio at Cincinnati.

1867 Joseph Monier patents "reinforced concrete".

1872 Birth of Robert Maillart, Swiss master of reinforced-concrete bridge design.

1874 James Eads builds his triple-arch bridge across the Mississippi at St. Louis: the first great steel bridge.

1876 USA's worst bridge disaster: collapse of Ashtabula Bridge, Ohio, on 29 December.

1879 Destruction of first Tay Bridge (opened 1977).
Birth of Freyssinet, French pioneer of concrete prestressing, and Ammann, US/Swiss designer of steel, and principally suspension, bridges.

1883 Construction of the record-breaking (486m/1,595ft) suspension Brooklyn Bridge by the Roeblings.

1884 Gustave Eiffel completes the Garabit Viaduct.

1886 Birth of Steinman, US bridge designer.

1889 Completion in steel of Forth Rail Bridge, the world's two longest spans (cantilever) each 521m (1,710ft).

1894 Tower Bridge, London.

1907 Collapse of unfinished Quebec bridge.
Birth of Fritz Leonhardt, German bridge designer.

1916 Hell Gate Bridge completed in New York – the first thousand-foot steel arch.

1917 Final completion of Quebec bridge – current world's longest cantilever span.

1929 Ambassador suspension bridge, Detroit, becomes longest span of any type at 564m (1,850ft).

1930 Maillart builds Salginatobel Bridge, Switzerland.
Freyssinet completes Plougastel Bridge, Brittany.

1931 The George Washington Bridge and Bayonne Bridge.

1932 Sydney Harbour Bridge.

1937 Golden Gate Bridge.

1940 Tacoma Narrows suspension bridge collapses.

1950 Freyssinet completes revolutionary prestressed-concrete bridges over the River Marne, France.

1951 Birth of Santiago Calatrava, Spanish architect and engineer.

1952-58 First modern cable-stayed bridges in Germany and Sweden.

1956 Construction of "world's longest bridge" (actually comprising hundreds of smaller spans) across Lake Pontchartrain, Louisiana.

1957 Then overall world's longest suspension bridge at Mackinac Straits.

1962 Lake Maracaibo Bridge, Venezuela: multiple cable-stayed and viaduct structure.

1964 New record suspension span, at Verrazano Narrows.
Europe's first long-span suspension bridge, Firth of Forth, Scotland.
Gladesville Bridge, Sydney: first thousand-foot concrete bridge.

1966 Tagus Bridge, Portugal: mainland Europe's first long-span suspension bridge.
Severn Bridge: first use of aerodynamic deck.

1970 Steel box-girder bridges collapse at Milford Haven, Wales, and Melbourne, Australia.

1978 New River Gorge Bridge, West Virginia: current world's longest steel arch.

1980 Ganter Bridge, Switzerland.
Longest concrete arch: Krk Island to Croatia.

1981 Then world's longest suspension span: Humber Bridge, England.

1985 T.Y. Lin proposes Inter-Continental Peace Bridge between Alaska and USSR across Bering Straits.

1987 "Image of the Bridge of the Future" competitions.

1988 First Honshu-Shikoku bridge complex, Japan.

1990 Feasibility study: Spain-Morocco crossing.

1991 Then world's longest cable-stayed span: Skarnsundet Bridge, Norway.

1992 Calatrava's Puente del Alamillo built at Expo '92.
First all-fibre composite bridge erected in Scotland.

1993 602m (1,975ft) cable-stayed bridge at Shanghai.

1994 Pont de Normandie: longest cable-stayed bridge.

1997 1,137m (4,519ft) double-deck Tsing Ma suspension bridge, Hong Kong, opened.

1998 Second Honshu-Shikoku link completed with Akashi-Kaikyo Bridge, world's longest span and longest overall suspension bridge; Great Belt complex, Denmark, completed with East Bridge, world's second-longest suspension bridge.

1999 Third Honshu-Shikoku link to be completed: includes Tatara Bridge, world's longest cable-stayed span.

2000 16km (10 mile) Øresund link between Copenhagen, Denmark, and Malmö, Sweden, completed; 1688m (5538ft) span Izmit Bay Bridge, Turkey, under construction.

2000> Strait of Messina Bridge, Italy; Femer Belt link (Denmark Germany); Pearl River Bridge (Hong Kong/China); Malacca Strait Bridge (Malaysia/Indonesia).

part one

The use of bridges is as old as mankind. Over millennia the simple hanging ropes of prehistoric times developed into elaborate suspended structures of bamboo and iron; crude stepping-stones were superseded by mighty pontoon bridges of cabled-together ships, and simple wooden spans were overshadowed by the mighty stone arches of Rome.

Crude but effective: jungle bridges such as this one of woven vines have served their purpose for tens of thousands of years.

1
ORIGINS

We cannot of course know when, where or how man first used a bridge, but many of our earliest hunter-gatherer ancestors must surely have had to range far and wide in their quests for food, fuel or shelter; and inevitably they must have encountered natural obstacles such as streams, rivers and chasms. To cross from one side to the other, for whatever purpose, casual use of what nature and chance threw their way had to come before any deliberate emplacement.

Where running water eats away soft strata or shale from beneath more durable layers of rock, natural stone arches can be formed, and perhaps bridges such as these were used by some of our forebears. Some of them were, and indeed are, enormous – for example, American Indian folklore tells of the mighty Tomanowos ("bridge built by the gods"), which once spanned the Columbia River between Washington and Oregon. It fell centuries ago, but the

Below **The Landscape Arch 89m (291ft) long and 32m (105ft) high, is one of the most spectacular of more than 1,000 natural rock arches in the Arches National Park, Utah. Its relative flatness may have made it of practical use for prehistoric native North Americans.**

rubble is said still to be visible on the river bed. Today, the longest natural rock spans still surviving are the Landscape Arch and the Rainbow Bridge in Utah, both spanning over 80m (260ft).

Infinitely more often, though, it must have been dead tree-trunks or branches, conveniently toppled across gaps, which formed the first bridges for the feet of our remote ancestors, ages before their descendants contrived to put in place what chance and nature had not provided. And how often did primitive man grasp a hanging jungle vine and swing out over a narrow rushing stream, unaware that millennia later the same fundamental interaction of materials and forces would be used to hang immense steel ribbons across incomparably vaster watercourses?

So whenever thought at last began to punctuate instinct, the models for the earliest bridge-erectors to follow had already been around for a very long time.

Right **The Tarr Steps across the River Barle, 6km (4 miles) northwest of Dulverton in Somerset, England. Did prehistoric humans once use these stones, which form one of the most elaborate of the ancient "clapper" bridges of southwest England? It has 17 granite spans, totalling 55m (180ft). The origin of the term "clapper" is now lost, although it may have been derived from the Anglo-Saxon, medieval Latin or old French words for "a pile of stones".**

Right, left to right **The simplest primitive bridges: a felled tree trunk; a crude deck of branches between edge trunks; a couple of stone slabs.**

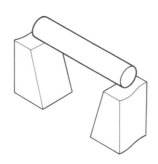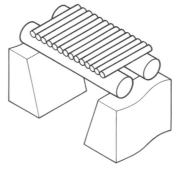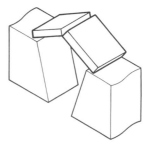

With the development of stone tools, crude at first but increasingly efficient as time passed, it was perhaps not too much of a collective mental leap for a particularly perceptive tribe to grasp the idea of hewing down a tree-trunk with flint hand-axes and dragging it across the narrowest width of an inconvenient stream, rather than search up and down the banks for shallows or wait for a log to fall into place of its own accord. It was then only a short step to felling several trees and laying their trunks side by side to make a broader surface, and then to lashing them to one another, so that the bridge held together. Then, rather than use whole trunks, why not split the timbers to form rough flat planks, both to make the bridge lighter and the crossing smoother? Downstream, perhaps, skills were progressing as far as laying smaller branches of a similar length transversely across the space between two long separate edge trunks, to form a wider and flatter bridge floor; while up in the hills, where perhaps faster-flowing waters were forced between narrow-sided gullies under overhanging rock ledges, others may have been bridging the gap that remained with a couple of slabs of rock propped against each other in an inverted "V", or, if it were long enough, with a single slab.

Sometimes the river would have been too wide to have been spanned by single beams, whether they were the timbers plentifully available in temperate parts of the world or the stone slabs of the harsher regions. In either case, however, if some conveniently flat rocks were at hand, and if the water was shallow enough, the rocks could be dragged in to form mid-stream stepping stones; and if the water was too deep for this, they could be piled up to make primitive piers, on which the ends of the beams could be balanced from the shore. Again, if the river-bed was suitably soft, it might have been more practicable to make the piers out of timbers or long stone slabs rammed in end-wise and buttressed at the sides, rather than crudely stacked.

Any such sequence of events must be speculative: these advances undoubtedly happened, but piecemeal. One thing is certain – the development of basic bridge-building skills was already well under way before the earliest era of recorded history.

Of course, no original prehistoric bridges are known for certain to have survived, but in many parts of the world there are still examples of types so basic that for all intents and purposes they can be regarded as such. In England, the timber "clam" bridges of south-east Cornwall and the stone "clapper" bridges of Dartmoor would seem to be examples of prehistoric construction technology – and who can be absolutely certain that some of these ancient granite slabs are not the very same stones as were trod by our remote forebears?

STRUCTURAL PRINCIPLES

Every bridge ever built, from the first log across a stream to the latest gravity-defying marvel, has used the properties of its materials to withstand, overcome and exploit the forces to which it is subjected. These forces have three origins: the bridge's own weight, the *dead load*; the traffic passing over it, the *live load*; and other external *environmental loads*, such as wind, earthquakes or water.

The proportionate amounts of stress imposed by each type of load can vary enormously according to the materials, the structural type, the bridge's use and its location, but within the fabric of all bridges (or any other structures), the actual nature of the

Below **A primitive beam bridge east of Qala Panji in Afghanistan. Such crude assemblages of branches and twigs are probably as old as *homo sapiens*.**

forces can always be reduced to four types, acting singly or in combination. Two are opposites: *tension*, which pulls apart, and *compression*, which pushes together. The others are *shear*, from shearing or cutting through, and *torsion*, or twisting. Theoretically, shear and torsion can be described in terms of tension and compression, but it is more practical to consider them as different forces.

The abilities of materials to cope with these forces are called their *strengths* (tensile strength, compressive strength, and so on). Of the materials available in pre-industrial times, stone has a good deal of compressive strength but is deficient in tensile strength, vines and ropes are the opposite, and different varieties of timber vary greatly in both – for example, balsa wood is not terribly useful for building durable bridges; but, compared with a stone beam, a stout trunk of hardwood, like oak, has much more tensile strength and not significantly less compressive strength. All these natural materials are vulnerable to shear, particularly vines and ropes, but while stone and timber are also vulnerable to torsion, vines and ropes have the flexibility to survive it. Irons and steels are generally superior in all four strengths.

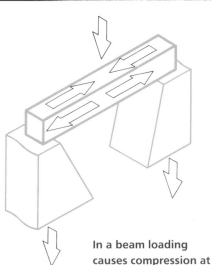

In a beam loading causes compression at the top and tension along the bottom, and bears downwards on the ground beneath.

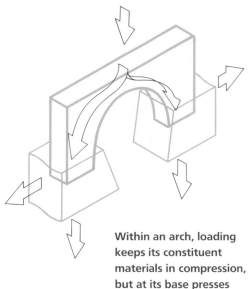

Within an arch, loading keeps its constituent materials in compression, but at its base presses downwards and outwards, necessitating abutments.

The combinations of different forces and the different strengths of materials produce varying effects in the basic types of bridge. On a simple *beam*, such as a log or a stone across a stream, the combination of its dead load and any live load results in *bending*, which means that, within its solid and, for the sake of argument, uniform, mass, it experiences a continuous *gradient* of stress, from purely compressive on its top edge to purely tensile along the bottom. (For simplicity this ignores any environmental load, which in practice might well add an element of torsion or even shear.) The tensile strength along the grain of a hardwood tree-trunk allows it to take a good deal of both forces before it breaks, but if a stone beam of similar proportions is subjected to the same loading, it will not bend to anything like the same degree before it cracks at its bottom edge.

In a beam, compressive and tensile forces are kept in balance. However, in the second basic bridge type, the *arch*, stresses are in compression, as dead and live load combine to press the material together. The arch form is thus ideal for construction in stone, because the material's compressive strength is used, while its lack of tensile strength becomes irrelevant.

The third basic bridge type, the *cantilever*, is really a development of the beam, rather than a separate form. In essence, it is a bracket secured at one end with the other end hanging free. The simplest form of cantilever bridge has a central beam supported between opposite cantilevers, but in practice the most characteristic design has several spans with cantilevers balanced evenly from one or more piers so that two or more spans are formed by the supported beams — so-called natural stone arches are in fact closer to being cantilevered.

In considering how the forces act within the basic bridge types, it is also important to consider how the structures interact with the ground. In beams the tensions and compressions from dead and live load balance each other within the structure, so that the only forces acting on the ground are vertically downwards. The same is true of centrally-supported cantilevers, but arches tend to push outwards as well as downwards, and so need secure abutments to prevent them from sagging and eventually collapsing.

Below left **A simple stone arch, the Packhorse Bridge at Wasdale Head, Cumbria, England.**
Below right **This precarious-looking wooden structure at Dudh Khosi, Nepal, clearly shows the cantilever principle. Such bridges can have either a cantilever at the other end, or a simple support for the deck.**

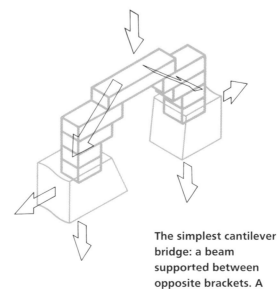

The simplest cantilever bridge: a beam supported between opposite brackets. A **certain amount of outward, as well as directly downwards, force can be generated.**

DEFYING GRAVITY: THE SUSPENSION PRINCIPLE

The beam and perhaps the cantilever, which must at some time have evolved from it, are not the only bridge types to have originated long before the dawn of history: different climates, terrains and materials led to different methods of spanning space. A single rope or cable across a chasm – in more analytical terms a *catenary* hanging freely between two points

of support – is also effectively a bridge. In fact, it is the most primitive form of another basic bridge type, the *suspension* bridge.

Today, the poised interaction of sky-piercing towers, taut cables and slender hanging deck constitutes probably the most potent image that the word "bridge" can conjure up; not really a solid, ground-based structure at all, but a leap into the unknown. Despite its apparently gravity-defying qualities, however, the suspension bridge still behaves in accordance with the same fundamental interactions of forces and material as the beam, cantilever and arch. Indeed, a suspension bridge can be regarded as an inverted arch with its properties reversed.

Primitive forms of suspension bridge are still in use all over the world. In remote parts of India there are examples as long as 200m (660ft) which consist of nothing more than a catenary of twisted bamboo rope. Travellers slide down it hanging from a bamboo loop, and when gravity-induced momentum gives out, they pull themselves along the remainder of the rope hand-over-hand. In a slightly more "user-friendly" version, they sit in a basket and pull themselves, or are pulled, across by means of a second cable; and in a version once found in China, travellers carried their own wooden saddle with a deep groove cut into the underside. On reaching a single-strand bridge, they fitted the cable into the groove, mounted and let gravity do the rest.

Among other primitive examples, one Chinese version simply adds a second rope a short distance

Above **The suspension bridge over the Tamba Kosi River near Kirantichap in Nepal is an interesting example of modern materials used in a form strongly reminiscent of primitive suspension bridges.**

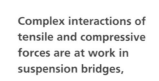

Facing page, engravings **The simplest forms of suspension bridge have been used from the earliest times to span distances over which any other type of structure would require far greater engineering resources.**
Facing page, main picture **This simple but perfectly serviceable suspension bridge of crude wire links crosses the Trisuli River in Nepal.**

Complex interactions of tensile and compressive forces are at work in suspension bridges, which must be designed to minimize their vulnerability to environmental loads.

above the first, which the traveller holds on to while walking across. Another, found in most parts of the world, has ropes on either side, like hand-rails. A further development has a more elaborate bottom and sides composed of many cables woven together transversely so that the cross-section becomes a "U", like a long, airborne hammock. In the 1930s, an expedition to the then Belgian Congo filmed the construction of such a bridge by a group of pygmies. First, a rope was suspended almost to the ground from the top of a tall tree overhanging the watercourse. Then a small boy clung to the hanging end and was swung further and further out until at last he was able to grasp a branch high up in a tree on the opposite bank. Once he had secured his rope, he clambered back and the first rope was then used to carry over others. After only six days, an elastic airborne half-tube, 50m (165ft) long, had been woven.

Inca constructions, some over 60m (200ft) long, had several massive cables, bound together, supporting a flat assembly of planks. Elsewhere – for example, in the Himalayas – primitive engineers have approximated even more closely the contemporary form, suspending their roadway from the cables.

These and many other examples show that early bridge builders used many varieties of the basic suspension type. But 2,500 years ago, skills had developed to such a degree that astonishing and ambitious bridges of quite different designs, were among the first artifacts to be described.

2

THE ANCIENT WORLD

For tens of thousands of years before the emergence of the first great civilizations, primitive bridges were simply used for aiding the movement of hunter/ gatherer tribes. Some 10,000 years ago, however, the rise of agriculture led to the first great change in society. Farming begat settled communities – villages, and then small city-states. Settlements needed buildings, farms required irrigation, and as populations expanded over the centuries, these needs engendered the development of engineering skills. Despite the achievements of the Egyptians and the Greeks in monumental architecture, the Roman civilization was the first to develop a real expertise in the design and construction of bridges. Even so, there are fragments of archaeological evidence, contemporary accounts and even rare standing structures which have survived from much earlier times.

Civil engineering began in Mesopotamia, where the Sumerians started digging canals for irrigation around 4000BC. Their early works were extended, developed and painstakingly maintained by successive peoples for some 5,000 years, to form perhaps the most extraordinary of all now-vanished engineering marvels. One small part was the network of 18 canals around Nineveh, constructed by Sennacherib early in the 7th century BC. To carry the waters of one canal to the city, he built a stone aqueduct, 280m (920ft) long and 20m (66ft) wide, across a small river-valley at Jerwan, 40km (25 miles) away. From the point of view of bridge construction, the interesting part was the 27m (90ft) section across the river itself, which was supported on five corbelled arches.

Corbelled arch   Voussoir arch

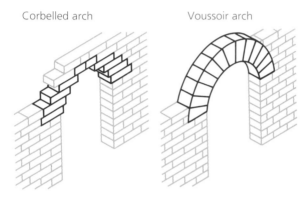

Right **The most celebrated use of the principle of corbelling in the ancient world was in the beehive tombs at Mycenae, Greece, c.1500BC. In the portico of the "Tomb of Clytemnestra", shown here, corbelling has been used two-dimensionally, but the concentric courses of masonry that form the domed interior represent its use also in three dimensions.**

Corbelling, the intermediate stage between a simple cantilever and a true arch, consists of successive courses of masonry placed on either side of an opening and projecting inwards closer and closer to each other until they meet. The principle was already known to the Sumerians and the Egyptians at least 2,000 years before Sennacherib's aqueduct. Surprisingly, the true arch form, constructed of *voussoirs* (stones cut into wedge shapes and placed in a semicircle) was also known in Egypt and Mesopotamia, apparently almost as early as corbelling.

In 626BC, the Chaldean leader Nabopolassar made Babylon his capital; and under his rule, and that of his son, Nebuchadrezzar II, it became the greatest metropolis the world had yet seen. A bridge across the Euphrates, built close to the stepped tower now identified as the Tower of Babel, was a central feature of the city. Estimates of its length vary from 120m to 200m (390-650ft); and excavations revealed the remains of seven piers in baked brick, stone and timber, each roughly 9m by 20m (30 x 65ft). The Greek historian Herodotus (c.490-425BC) ascribed the construction of what may be the same bridge to a Queen Nitocris, who he said diverted the river into an artificial basin and built the bridge in the dry: "As near as possible to the centre of the city, she built a bridge over the river with the blocks of stone which she had prepared, using iron and lead to bind the blocks together. Between the piers of the bridge, she had squared baulks of timber laid down for the inhabitants to cross by – but only during daylight, for every night the timber was removed[. . .] Finally, when the basin had been filled and the bridge finished, the river was brought back into its original bed[. . .]." (Were the timbers really taken away? Or was Herodotus describing the world's first bascule, or lifting, bridge?)

Herodotus has also left us accounts of three important military bridges. At the end of the 6th century BC the Persian monarch Darius ordered floating pontoon bridges to be built across both the Danube, close to its mouth in the Black Sea, and the Bosphorus. The most celebrated military construction, however, was the one which Darius's son Xerxes built in 480BC across an even more formidable obstacle, the Hellespont – a distance of over 1.5km (almost a mile).

Two rows of ships were said to have been lashed together, 360 on the Black Sea side, and 314 on the other, "moored slantwise to the Black Sea and at right angles to the Hellespont, in order to lessen the strain on the cables". A first pair of bridges, using flax and papyrus cables respectively, had already been destroyed by a storm. As a result, "this time, the two sorts of cables were not used separately for each bridge, but both bridges had two flax cables and four

A 19th century engraving of an unknown, perhaps Indian, pontoon bridge. Herodotus' account of Xerxes' vast assembly built across the Hellespont in 480BC indicates a double row totalling nearly 50 times the number of ships shown here.

papyrus ones[. . .] The next operation was to cut planks equal in length to the width of the floats, lay them edge to edge over the taut cables, and then bind them together on their upper surface. That done, brushwood was put on top and spread evenly, with a layer of soil, trodden hard, over all. Finally, a paling was constructed along each side. After due ceremony Xerxes directed his regiments across the bridges; so vast were their numbers that it was said seven days and nights passed before all had marched over.

From perhaps as long ago as 2000BC, the ancient Chinese built pontoon bridges, made out of sampans moored a few feet apart parallel with the current – very different from Xerxes' mighty assembly.

ROMAN FOUNDATIONS

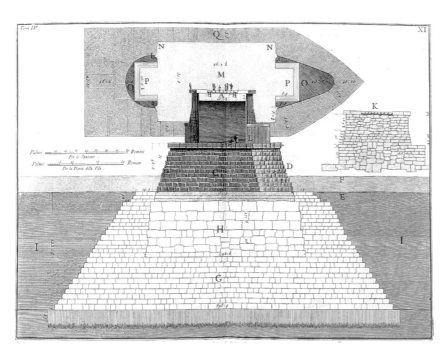

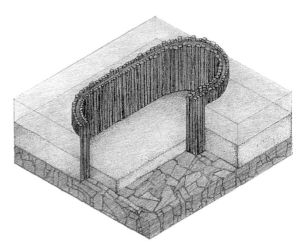

"With weeping and with laughter
Still is the story told,
How well Horatius kept the bridge
In the brave days of old."
  Macaulay: *Lays of Ancient Rome*

The first bridge in Rome of which we know the name was the Pons Sublicius, said to have been the one defended against the Etruscans in the 6th century BC by the legendary hero Horatius Cocles. *Sublica* means "pile" or "stake", but it is probable that, as on most Roman bridges, timber was used for the entire structure and not just the foundations. The popular conception of the Roman bridge as a series of monumental stone arches derives simply from the fact that so many stone bridges have survived, whereas every single timber bridge has perished. One does survive in reports, however: in his *De Bello Gallico*, Julius Caesar gives a remarkably detailed account of a timber trestle bridge, perhaps 400m (¼mile) long, which he ordered to be built across the River Rhine near present-day Coblenz in 55BC.

By the 2nd century BC Roman bridge engineers had mastered the techniques of both creating secure mid-stream foundations and constructing masonry works above. Three factors were crucial in this. Firstly, and all-important in bridge foundations, they discovered an excellent waterproof cement, made by mixing water, lime and sand with a fine powder ground from the volcanic rock, or tuff, that was found in large quantities near the town of Pozzuoli

Above left **Scale drawing by Piranesi of the central pier of the Ponte Sant'Angelo in Rome. The base, founded on timber piling, is 70m (230ft) long. The plan at the top is an enlarged view of the upper part of the pier, protruding from the water.**

Above right **The principle of one kind of Roman cofferdam. A double ring of wooden stakes was driven into the riverbed around the planned location of a bridge pier by a manually operated pile-driver. Clay was packed into the division between the two circles, and then the water was emptied from the enclosed space. After this, mass concrete or timber pile foundations were installed, and the bridge pier built on top.**

(hence its name *pozzolana*). Secondly, they developed a method of constructing foundations within a temporary enclosure called a *cofferdam*. Thirdly, they realized that the voussoir arch could be made to span much greater distances than any unsupported stone beam, and that it was stronger, more secure and durable than any other structure that could be built with the materials available at the time.

Roman arches were semicircular, which had important consequences. In a semicircular arch, more of the thrust goes directly downwards than in a shallower arch, which in consequence requires strong side-bracing at the abutments. This meant that, if the piers were wide enough (one-third of the span), any two could support a complete arch without further propping from the sides. Thus, it was possible for Roman engineers to build their bridges out from the shore a span at a time – cofferdam, then foundations, then pier, then arch – rather than having to go through the much more difficult operation of putting the entire bridge sub-structure in position first.

The world's seminal textbook on building, *De Architectura*, written towards the end of the 1st century BC by the Roman architect and military engineer Vitruvius, contains no section specifically devoted to bridge-building, but does describe cofferdam construction in the context of harbourworks: "Then, in the place previously determined, a cofferdam, with its sides formed of oaken stakes with ties between them, is to be driven down into the water and firmly propped there; then, the lower surface

Right **An engraving by Piranesi depicting four of the five arches of the Ponte Sant'Angelo. Although the piers above water level and the superstructure of the bridge are massive, the illustration on the previous page shows how they are dwarfed by the immense substructure of the bridge.**

inside, under the water, must be levelled off and dredged, working from beams laid across; and finally, concrete from the mortar trough [. . .] must be heaped up until the empty space which was within the cofferdam is filled up by the wall.''

Vitruvius goes on to describe the procedure if raw materials for concrete are not available: ''A cofferdam with double sides, composed of charred stakes fastened together with ties, should be constructed in the appointed place, and clay in wicker baskets made of swamp rushes should be packed in among the props [. . .] let the space now bounded by the enclosure be emptied and dried. Then, dig out the bottom within the enclosure. If it proves to be of earth, it must be cleared out and dried till you come to solid bottom and for a space wider than the wall which is to be built upon it, and then filled in with masonry

Above **The Ponte Sant'Angelo still proudly spans the Tiber, over 1,850 years after its completion. The present balustrade, which includes a cast-iron railing, and the statues of angels above the pilers, was added in the 17th century by Bernini.**

consisting of rubble, lime, and sand. But if the place proves to be soft, the bottom must be staked with piles made of charred alder or olive wood [. . .].''

Sometimes, a timber framework was constructed on land, manoeuvred into position and then sunk to be filled with the material of the pier. Whatever the method, the dangers were great for the Roman labourers, who often worked in fast-flowing currents with only crude, hand-operated machinery. Unsurprisingly, there were many times when the piling did not go deep enough for long-term stability; nevertheless, enough Roman bridge foundations were constructed with sufficient expertise for some bridges to have survived as usable structures for 2,000 years.

A consequence of needing massive piers to support the semicircular arches was that the combined width of these piers very considerably reduced the width of the channel, thereby increasing the speed of the current and concentrating its flow past and around them. This in turn increased the effects of scour and swirl on the piers; to counter this the Roman builders adopted an elongated shape, a cutwater, which was pointed to cleave the water more effectively.

With the piers complete, construction of the arches alone remained. For this a wooden formwork, or *centering*, was built out from the piers and braced to them, with the upper surface of the timbers shaped to exactly the required semicircular profile. Parallel arcs of stones were then placed on the centering behind each other to create the arch. The semicircular shape meant that all the voussoir stones could be cut identically, and also that no mortar was necessary to bind them: once the keystone in the centre was positioned, the compression forces, together with the perfect cut and fit of the blocks, ensured complete stability.

ROMAN AQUEDUCTS

The aqueducts which carried water into the cities of the Roman Empire were often extraordinarily extensive. Rome itself was served by a network of 11, which, on completion of the last, in AD226, totalled 560km (348 miles) in length. Only one seventh of that distance was above ground, but of that, as much as 60km (37 miles) had to be carried on arches to keep it high enough to maintain flow. Elsewhere, the systems were smaller than in the capital, but on occasion they required structures of extraordinary height and length in order to carry the water across valleys. Indeed, the Romans' greatest single aqueduct, built in the reign of Hadrian to serve Carthage in North Africa, extended no less than 141km (87½ miles).

The two most impressive to have survived more or less intact are the Pont du Gard, near Nîmes in southern France, and the one at Segovia in central Spain. The latter, at over 800m (2,624ft), is the longer and has a double row of 5m (16½ft) arches, one above the other, reaching to 36m (118ft) at the highest point. It was built on the orders of the Emperor Trajan at the beginning of the 2nd century AD as part of a system to bring water to Segovia from the Guadarrama Mountains, l00km (62 miles) distant.

The Pont du Gard is, if anything, even more monumentally striking. The Swiss philosopher Jean Jacques Rousseau, pacing beneath the immense arches, "felt lost like an insect in the immensity of the work". It is thought to date from about a century earlier than the Segovian structure and was probably built by the Emperor Agrippa in around 19BC as part of a 40km (25-mile) aqueduct; 270m (886ft) long, it carried the water across the River Gard, a tributary of the Rhone, at a maximum height of no

Below **Apart from its sheer scale, the most striking feature of the Segovia aqueduct in central Spain, is its slenderness. Even the bases of the piers are only 2.4 m (8ft) wide.**

less than 49m (160¾ft) – taller than the nave of any Gothic cathedral. Here, there are three tiers of arches. The first two consist of broad spans symmetrically arranged one above the other, but as the sloping sides of the valley require them to cover a much greater overall distance at the higher level, there are 11 on the second tier and only six on the first. All of them are exceptionally broad, even by Roman standards. The longest, which crosses the river itself, is 24.5m (80½ft), and the others vary between 15.5m (51ft) and 19.2m (63ft). The unknown designer daringly departed from the 3:1 pier/span ratio that was common practice in Roman bridge-building; at the Pont du Gard it is nearer 5:1. On both the lower tiers, the ends of some stones have been left proud of the surfaces of the piers and spandrels, probably in order to support scaffolding.

The water was conveyed in a deep, cement-lined channel above the third tier, which consists of 35 uniform 3.5m (11½ft) arches. The piers here are virtually as broad as the spans, which gives the impression of a continuous masonry wall punctuated by semi-circular openings; and the effect of this seeming massiveness supported by the soaring grace of the lower tiers is not only overwhelmingly beautiful but also an eloquent testimony to the natural load-carrying capacity of the voussoir arch and the skill of the masons who cut and shaped its blocks. For 20 centuries they have stood without benefit of mortar, although the Pont du Gard has intermittently suffered considerable damage. In 1747 the width of the lowest tier was doubled when a roadway was added alongside it with arches that exactly matched the Roman original.

Left and below **The great Pont du Gard near Avignon, arguably the most beautiful of all surviving Roman engineering works. In the photograph below the projecting stones left for scaffolding can be clearly seen. Mortar was used only in the third tier.**

Previous page, top **When we think of Roman bridges today, we tend to think of stone arches, but far more were constructed of wood. All are long since destroyed. Trajan's Column in Rome bears a bas-relief depicting the mightiest of them, the 21-span, near 1,000m (3300ft) long structure built on 20 towering stone piers across the Danube in the province of Dacia (now in Romania) in c. AD100.**

TIMELESS ARCHES: ROME'S GREATEST STONE BRIDGES

Roman masonry bridges, both fragmentary and in rare instances complete, have survived in most of the countries that were once part of the Empire. It is not surprising that Rome itself has more enduring examples from its Imperial hey-day than has anywhere else. Of eight known to have existed, as many as six can still be seen, although four of these – the Ponte Rotto (2nd century BC), the Ponte Molle or Pons Milvius (110BC), the Ponte Cestius (43BC) and the Ponte Sisto (AD370) – remain only as whole or part-spans, either standing alone in ruin, like the Ponte Rotto's one surviving arch, or incorporated into later restoration, like the Pons Milvius's pair. (It is salutary to reflect that the time-span between the first and last of these four bridges corresponds to that between the mid-15th century and the present day.)

The other two bridges, the Ponte Quattro Capi or Pons Fabricius (62BC) and the Ponte Sant'Angelo, originally the Pons Aelius (AD134), survive as recognizable Roman bridges, though with later decorations and alterations. The five arches of the Pons Aelius, spanning the whole River Tiber, form a more imposing whole than the two arches of the Pons

Below **The Pons Fabricius (built in 62BC), one of Rome's two most perfectly preserved ancient bridges. At over 27.4m (90ft) each, its two arches are exceptionally long-spanned by surviving Roman standards. The small weight-relieving central arch was matched by others that are now buried in the more recently built abutments.**

Fabricius, which simply link the Tiber's left bank with the small island of Aesculapius; but the structures of both bridges incorporate an important feature which is also found on some other Roman bridges – small additional arches in the spandrels above their central piers and in the abutments, designed to reduce weight and provide additional channels for floodwaters. At first, these were thought to exist only on the Pons Fabricius, but in the late 19th century, during restoration and rebuilding, similar spans were found on the Pons Aelius as well.

Two further bridges, neither of them in Rome, exemplify complementary aspects of the Roman legacy better than any others. The first is the Pons Augustus at Rimini, built before AD14 and thus contemporary with the Pont du Gard. Urban and modest in scale – its five spans are between 5.1m (16¾ft) and 4.2m (13¾ft) – it is nevertheless elegantly decorated with niches on the spandrels over the piers and sharply-etched dentils (small, squared supports, like spaced teeth, carrying the cornice). This was the bridge which, 1,500 years later, Andrea Palladio adopted as his most important model; and in consequence it was

Right **The masterpiece of Caius Julius Lacer, the Puente de Alcántara, in Caceres, Spain. Such a structure, so far from the Imperial centre, remains an eternal testimony to the power and influence of Rome in its heyday.**

to have an extraordinarily widespread influence on bridge design thereafter. As the enthusiasm for Classicism continued throughout the ensuing centuries, many different architects designed bridges in the "Palladian" style.

Perhaps the greatest of all Roman stone bridges is in complete contrast to the Classical, travertine limestone-faced sophistication of these bridges. Lofty, remote, strategic and awe-inspiring, the Puente de Alcántara in Spain crosses a steep, narrow valley of the Tagus River, close to the border with Portugal. Everything about it speaks in superlatives, from the towering granite piers up through the six great arches – the central two of which are the longest to have survived from Roman times – to the parapeted roadway more than 50m (164ft) above the riverbed, higher even than the Pont du Gard. As on that other masterpiece, the granite voussoirs were laid without mortar, but have nonetheless survived the floods of nearly 2,000 years.

Like the timber bridge in Dacia, and indeed the Segovian aqueduct, the Puente de Alcántara was the work of the Emperor Trajan, but here the name of the engineer himself is also known. Nearby is the tomb of Caius Julius Lacer. Who he was and how he came to die beside his creation are not recorded, but the words of his epitaph can stand as a truthful testimony not only to his bridge but to the enduring influence of the Empire he served: "I leave a bridge forever to the generations of the world".

THE PUENTE DE ALCÁNTARA FACTS	
constructed	**c. AD100**
total length	**180m/600ft**
max. span	**30m/98½ft**
breadth of main piers	**9m/30ft**
span:pier ratio	**3.3:1**
max. height	**50m/164ft**

3

"SUR LE PONT D'AVIGNON"

The Pont d'Avignon in southern France was at once the largest, almost the earliest, and structurally one of the most daring and sophisticated bridges of its time. The collapse of the Roman Empire in the West brought to an end a long era of engineering achievement, including the construction of bridges with any pretensions to permanence. It was only with the building of the Pont d'Avignon, which began in 1179, that Western medieval engineers for the first time produced a bridge that matched the masterpieces of the Romans.

It was not so elsewhere. In Persia and China in particular, many notable masonry bridges were built before the 12th century AD. In AD260, before the fall of the Roman Empire, and as a consequence of one of its rare defeats, the Persian ruler Shapor I employed Roman captives to construct a major bridge over the Karun River in Persia. Using Roman methods, they erected 40 arches with a total length of over 500m (1,640ft). One aspect of Roman expertise not adopted here, however, was the use of cofferdams. Instead, the line of the bridge meandered across the river using stone outcrops wherever possible as the bases for the piers. As a result, some of the piers were so massive that to a certain extent the bridge doubled as a dam. Other bridges built by the Persians used pointed, or ogival, arches, probably because they were somewhat lighter and required

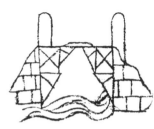

Above **A diagram entitled "How to make a bridge over water with twenty-foot timber" from a page on Practical Geometry in the sketchbook of the 13th century French engineer Villard de Honnecourt. This clever cantilever truss design was probably a response to the depletion of timber in France, and Europe generally, following increased demands for wood for fuel, building and other industrial purposes. This made wooden beams for bridge-building longer than "twenty-foot" scarce.**

Left **A semicircular archway at the ruins of Kirkham Priory, Yorkshire, England. This is a typical feature of "Romanesque" architecture, which flourished in Europe between the 10th and 12th centuries, and which reflected the Roman experience in creating semicircular vaulting and arches for bridges and buildings.**

less massive centering – an important consideration in a country with very little timber.

In China also, bridge-building flourished while it stagnated in Europe. The semicircular masonry arch form had spread eastwards from Rome and had probably influenced Chinese designs, but with one bridge the Chinese engineers executed a structural leap forward that was not to be matched in the West for seven centuries. This was the An Ji Bridge at Zhao Xian in Hebei Province. A span of 37m (121ft) for a stone arch bridge completed in AD605 is striking enough, but far more so are two further design features. First, the haunches on both sides of the bridge are pierced by pairs of smaller arches, in order to lighten pressure on the abutments – similar to but more advanced than the Roman practice. Secondly, and most remarkably, the main arch itself is segmental: the arc-section that forms the span is only a small part of a circle, in contrast to the whole semicircle used by the Romans. As a result, the arch of the An Ji Bridge rises to a height of only 7m (23ft), despite its length, and in consequence the rise of the roadway is unusually gentle. On most Chinese bridges, the often remarkably slender crowns of their semicircular arches – which were flexible, unlike those of the Romans – soared high above the riverbanks, necessitating flights of steps on both sides. By the 13th century, the Venetian explorer Marco Polo was able to report the existence of thousands of these bridges – a testimony to what must already have been centuries of expertise.

Elsewhere in China, hugely different bridge designs had been executed. For example, the principle of cantilevering was taken to one logical conclusion

in the 10th century on the Poh Lam bridge over the Dragon river in Fukien. The structure of this bridge consists of colossal corbelled granite beams weighing perhaps as much as 200 tons each and providing spans of up to 20m (65½ft).

At the opposite extreme from such monumentality were the iron-chain suspension structures which travellers found in China in the eighth century and even earlier in India. Remarkably, a chain-suspended bridge also existed in early medieval Europe – the Twärrenbrücke, or Transverse bridge, which was erected in the Schöllenen Canyon in around 1218 in what became Switzerland. It was an isolated example, perhaps, but from the 11th century onwards European technology – particularly in mining, agriculture, milling and construction – expanded at a rate that is still little appreciated. One seminal source is the sketchbook of the 13th-century architect and engineer Villard de Honnecourt, whose designs include a timber bridge and a device for sawing off the tops of timber piles under water.

In the Middle Ages the semicircular Roman arch, which had continued as a Romanesque tradition in church-building, was superseded by the pointed arch. Though it is of great antiquity – examples in Mesopotamia have been dated to at least the second millennium BC – the pointed arch's full structural possibilities were not explored until the great Gothic Age of cathedral-building, beginning in the mid-12th century. Pointed arches were not merely used as door and window openings, they also carried structural weight from ribbed vaulting at roof level down through often immensely tall columns, buttressed on the outside. This released the walls from carrying the major load of the roof and as a result opened them up to glorious expanses of stained glass.

Not surprisingly, pointed arches were widely used in medieval bridge building, at least for smaller

Right **An archetypal Gothic pointed arch at St Osyth Priory, Essex, England. This is a fairly late and elegant example, dating from the 15th century, of the style which began to appear In Western architecture from about AD1150 onwards. Pointed arches had two diverse uses: for churches and cathedrals in particular, they enabled a mighty extension to height, openness and grandeur; in bridge-building, they were a way of constructing relatively small spans with less expertise than was needed for semicircular arches. Consequently, medieval brldges which incorporated pointed arches, such as Old Bristol Bridge and Old London Bridge, tended to be as crudely constructed, workaday and ramshackle as their ecclesiastical neighbours were soaring monuments of sublimity.**

spans. They were easier to construct than semicircular arches, as their voussoirs did not have to be as precisely matched and dressed for them to be reasonably stable. A case in point was Old London Bridge, which was begun in 1177 by a monk named Peter of Colechurch on a site only 300m (900ft) wide, which had seen the successive building and destruction of wooden bridges since Roman times. This, the most famous of all medieval bridges, took 33 years to complete. By the time its 19 small, ill-matched, pointed arches were finished, spanning between equally irregular boat-shaped piers or "starlings", its achievement had been completely eclipsed by the magnificent structure in southern France.

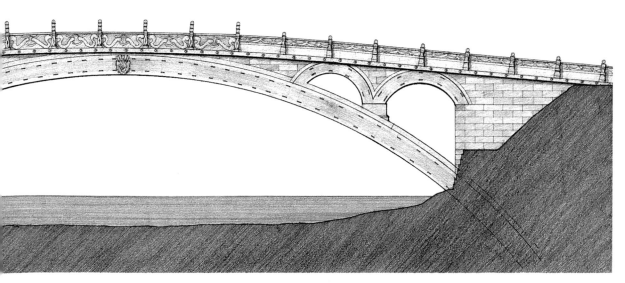

Left **The An Ji Bridge in Zhao Xian, Hebei Province, China, built during the Sui dynasty, in the 7th century AD, is one of the most remarkable stone bridges in the world, predatlng anything comparable in the West by 700 years. The extremely shallow segmental arch (28 courses of masonry span the river) and the open spandrels were unprecedented.**

LES FRÈRES DU PONT: THE RHÔNE

FRANCE

The extensive ecclesiastical building programme which accompanied French religious fervour at the end of the 12th century seems to have included bridges as well as cathedrals and monasteries. The provision of a bridge was regarded as an act of pious charity, and specially dedicated orders were established to build and maintain masonry bridges and hospices at dangerous crossings.

According to Viollet-le-Duc, the 19th century architectural historian, the first of these bridges was built over the River Durance, at a place originally known as Maupas and subsequently renamed Bonpas, by a monk called Bénoît in 1164. In the years that followed, several other bridges were erected under Brother Bénoît's direction by the group of monks who later became known as "Les Frères du Pont"; and Viollet-le-Duc also credits him with the famous bridge over the River Rhône not far from Bonpas, at Avignon. But local legend, and indeed the Catholic Church, ascribe the bridge to a shepherd called Bénèzet, who had a vision in 1178, in which he

was commanded by God to build it. When the Bishop of Avignon demanded evidence of his divine inspiration, Bénèzet miraculously lifted a massive stone and carried it to the place on the river-bank where the bridge was to begin.

Whatever the truth of its origins, the Pont d'Avignon was a tremendous achievement. The total length was about 900m (3,000ft), stretching in a long straight line across the Petit-Rhône, over the island of Barthelasse and out into the middle of the Grand-Rhône, where it turned sharply towards the west bank in a "V" shape, which seems to have been designed to enable the bridge to withstand the very high spring floods. Apart from its sheer scale, the bridge exhibited structural mastery amounting to genius. It had only one or two more spans than the exactly contemporary London Bridge, and yet it was three times as long. All of the four spans that have survived the ravages of wars and ice are longer than any remaining Roman arch; and although they may incorporate some 14th century reconstruction, their form

Below **All that is left of the Pont d'Avignon, the longest bridge of medieval times – four spans each more than 30m (100ft) out of the original 20 or 21. St Bénèzet's chapel stands over the second pier. Small relieving arches are incorporated above the piers and in the spandrels to accommodate spring floodwaters.**

remains masterly and innovative. Instead of being semicircular or pointed, they were *elliptical* – the curvature of the arc, while remaining rounded, becomes much tighter towards the apex. The result of this was that the piers could be made narrower than on other bridges and the arches taller than semi-circular arches of the same span, thereby carrying the roadway higher out of the way of potential flooding. For its length, the bridge was remarkably narrow, reducing from a maximum width of 5m (16½ft) to only 2m (6½ft) by the side of the chapel, which still stands atop the first pier; and the sense of slender elegance which this created was accentuated by the remarkably short distance between the roadway and the crown of each arch. Although most medieval bridges were made with mortar, the voussoirs here

Above **Built in the early 14th century, some 150 years after the Pont d'Avignon, the Pont Valentré over the River Lot near Cahors in southern France is perhaps the most beautiful and perfectly preserved of all French medieval fortified bridges. A third defensive tower stands behind the trees on the left. The six arches each span 16.5 m (54 ft).**

were so finely dressed that they could be laid without it. Perhaps most remarkably of all, the whole magnificent structure was built in only ten years.

The supposed creator of the Pont d'Avignon did not live to see his masterpiece completed. He died in 1184, but he was later canonized, and today the body of Saint Bénèzet still lies interred in the chapel, which has become a place of pilgrimage.

PONT D'AVIGNON	FACTS
originally constructed	**1178-87**
original no. of spans	**20-21**
original min./max. spans	**approx. 20m/ 65ft to 35m/115ft**
surviving min./max. spans	**30.8m/101ft to 33.5m/110ft**

THE PONTE VECCHIO, FLORENCE
ITALY

"Taddeo Gaddi built me; I am old,
Five centuries old."

Longfellow

Medieval Florence was a flourishing commercial capital. The links that led to southern cities across its flood-prone River Arno were as essential to its continued prosperity as was the success of its bankers and its weavers. Timber bridges had been built both by the Romans and during early medieval times. A Ponte Vecchio was already in existence in 1077, when the Florentines decided to bring the course of the river within the city walls, but this bridge fell victim to flood in 1177, and so too did a successor, in 1333. This further disaster demolished other bridges as well, and it caused so much additional destruction that the citizens undertook a long and extensive re-

Below **The houses and shops which remain to this day along the length of the Ponte Vecchio are a rare example of such medieval use of bridge-space. The original butchers' shops were replaced in the 16th century by jewellers'.**

construction programme, building up the river's embankments in an attempt to make the nearby areas of the city as safe from further catastrophe as possible. As a result, work did not begin on a new Ponte Vecchio until 1345.

The design of the Ponte Vecchio that we know today has been attributed to Taddeo Gaddi (c.1300-1366), a pupil of the painter Giotto. To cross the 100m (330ft) flow of the Arno, Gaddi erected his bridge on a symmetrical arrangement of three arches. Although the building of this Florentine bridge can no longer be regarded as the first use of the segmental arch, it was nevertheless a radically important innovation in terms of European architecture. Whether it was Gaddi or in fact some other architect who realized that it was possible to depart with advantage from the Roman semicircular model,

he could hardly have known about the An Ji bridge, which had been built in China centuries earlier (see pp. 26-7). The design of the Ponte Vecchio was a step into the unknown.

The Ponte Vecchio remains a remarkably daring structure – its arches are extremely shallow compared with any previous, or indeed contemporary, European bridge. The rise of the arches varies from 3.9m (12¾ft) to 4.4m (14½ft), giving a span/rise ratio from water level to the soffits of only about 5:1. The piers, on the other hand, were made very thick, which was in line with the traditional Roman practice and was doubtless also regarded as a precaution against the then unknown and incalculable horizontal thrusts that would be generated by low segmental arches.

These arches provide both longer spans and a lower roadway than semicircular arches, which was doubly advantageous in the case of the Ponte Vecchio. Fewer piers meant less resistance to floodwaters, and the reduced height of the bridge-deck, which was twice as wide as its predecessor, later allowed the construction of the two-storey structure along the length of the bridge which gives it its characteristic and world-famous profile – an upper gallery linking the Pitti and Uffizi Palaces, and a lower level flanked by a double row of jewellers' shops. Other Florentine bridges continued to be sus-

Above **The influence of this small city in art, architecture, commerce, economics and politics stretched far beyond its boundaries, and indeed those of the country of which it was once the capital. In 1564-5 a vaulted walkway was built across the Ponte Vecchio to link the ruler's palace with the new government offices. Now the former, the Pitti, and the latter, the Uffizi, are two of the most richly endowed art galleries in Florence.**

ceptible to flooding of the Arno, but the Ponte Vecchio has survived to this day, escaping even the ravages of the Second World War.

The potential for long-span construction inherent in the segmental arch was spectacularly realized within a few years of the Ponte Vecchio's completion. The Ponte Castelvecchio, built between 1354 and 1356, extended nearly 50m (164ft) across the Adige River at Verona; and in 1370 and 1371 the Duke of Milan constructed a bridge with a clear span of no less than 72m (236ft) across the River Adda at Trezzo, to provide access to his castle. This record-breaking arch, which was twice the span of any constructed by the Romans, was not to be surpassed for four centuries, and has been exceeded very rarely since then in masonry construction. Unfortunately, however, this impressive structure was destroyed in 1416, during a siege of the castle.

THE PONTE VECCHIO	FACTS
main span	30m/100ft
side spans	27m/90ft
thickness of piers	6m/20ft
span: pier ratio	5:1
total width	32m/105ft
rise of arches	3.9m/12¾ft
	to 4.4m/14½ft

THE CHARLES BRIDGE, PRAGUE
CZECH REPUBLIC

As the medieval centuries progressed, the growth of industrial and commercial activity in many European cities generated a volume of traffic which could only be accommodated by durable masonry bridges that were also capable of withstanding floods. Or, in terms of structural properties, a growth in the magnitude of live loads, combined with intermittent extremes of environmental load, necessitated the building of structures which could shrug off not only increased compressive forces from above but also shearing and torsion from the sides and below.

A case in point was the city of Prague, which is divided by over half a kilometre of the swiftly flowing River Moldau or Vltava. A wooden bridge had been constructed over the river at the beginning of the 12th century, but it had been destroyed by flood in the middle of the same century and subsequently replaced by a stone structure, known as "the Judith Bridge" after the wife of the reigning King, Vladislav I. Completed by 1172, the Judith Bridge must have been a tremendous engineering achievement for its day. It was only the second medieval stone bridge in Central Europe, after that at Regensburg in Germany (which is still standing). The Judith Bridge survived for 170 years, until it was destroyed by another flood in 1342.

Above **Tourists constantly throng the Charles Bridge, now open only to pedestrians. Beyond is the May Day Bridge, completed in 1901. This replaced an early 19th century chain suspension bridge, which had been the first to span the river after the Charles Bridge. In the distance are the Jiráskův and Palackého Bridges.**
Facing page, main picture **The Old Town Bridge Tower, on the right bank, was begun at the end of the 14th century to a design by Peter Parler. It was started under Charles IV and continued under Václav IV.**

The replacement, the present Charles Bridge, was begun in 1357 by the great King Charles I, who, as Charles IV, had also been crowned Holy Roman Emperor (although the bridge which now bears his name was not actually called after him until 1870). Work on the bridge was begun under the supervision of a 27-year-old architect, Peter Parler of Gmünd. The founding of the piers turned out to be a lengthy and difficult procedure, but according to Czech sources, the new bridge was essentially completed and in use by the 1380s, although some accounts claim that its actual construction extended over almost a century and a half. In a sense the claim is true, as successive floods and a consequent weakening of the piers necessitated intermittent strengthening work throughout the ensuing century, most seriously in 1432, when the bridge was badly damaged in three different places. It was constructed in sandstone which is much softer than granite.

The Charles Bridge incorporated no notable structural innovation, but its scale makes it arguably the most monumental bridge of its time. The total length of 515.76m (1,692ft) is carried by 16 arches, which vary in span between 16.62m (54½ft) and 23.38m (76¾ft), and which support the 10m (33ft) wide deck in a slight zig-zag across the longest continuous

Inset **The Charles Bridge has 15 statues or groups on each side. After the first, St John of Nepomuk in 1683, a** *pietà* **was added in 1695-6, followed by 26 more between 1706 and 1714, including in 1709 SS Saviour** (far left), **and Cosmas** (near left). **Finally, St Christopher and St Wenceslas were added in 1857 and 1859 respectively. St John is in bronze, all the others are stone. Some necessary replacements have been made over the years.**

extent of water covered by any medieval bridge (the Pont d'Avignon was partially on an island). The body of the bridge, particularly the wedge-shaped piers, which range in thickness from 8.5m (28ft) to 11m (36ft), was made extraordinarily massive in order to withstand the floating ice of winter.

As it was continuously repaired, modified and enhanced over a very long period, the Charles Bridge can be seen as a link between the styles of different ages. Its arches, with their near semicircles, follow the traditional Roman manner, whilst the piers are typically medieval. The Bridge terminates on the left bank in two towers, the "lower" of which dates back to the 12th century and formed part of the fortifications of the Judith Bridge. In 1590 it was renovated in Renaissance style by which time its "higher" companion had been built to match in style the Old Town Tower at the opposite end of the bridge. Thus the towers between them incorporate four centuries of stylistic evolution from early Gothic to Renaissance. This architectural cocktail was further enriched when the embellishment of the parapets with Baroque devotional statuary began in 1683. The first was St John of Nepomuk, who had been thrown from the bridge and drowned in 1393 at the order of Václav IV, following a religious dispute.

4

RENAISSANCE TO ENLIGHTENMENT

Paris joined the metropolitan stone bridge-builders at the beginning of the 16th century, with the construction in only seven years (1500-07) of the six-arched, 124m (407ft) Pont Notre-Dame between the right bank of the Seine and the Île de la Cité. This replaced a timber bridge which had collapsed in 1499, and like it, carried a double row of houses on its 23m (75½ft) width. It was designed by an Italian, Fra Giovanni Giocondo, who is also credited with the introduction of a feature sometimes known as the *corne de vache*. This involved splaying out the arch by chamfering its edge (or *archivolt*), which not only added elegance but also allowed a broader deck above and aided the passage of flood waters.

This feature was used on the left-bank arm of the next Parisian bridge, the Pont Neuf (1578-1607). Although it survives to this day, albeit with arches flattened from semicircular to elliptical during l9th century reconstruction, the Pont Neuf is a triumph of irregularity. All 12 arches differ in span, not only from each other, but also between their upstream and downstream sides.

The most imposing French bridge of the Renaissance was another "Pont Neuf", this time in Toulouse, which was completed in 1632 after almost a century of building. It boasted an exceptionally long 31.7m (104ft) main span, with an arch that was slightly flattened from a true semicircle. However, the most remarkable evolution of the curve in a Renaissance bridge was to be found in the heartland of the movement, in Florence, on the Ponte Santa Tri-

Right **Elevations of two of Palladio's examples of truss design. The upper one is remarkably similar to the structure which Lewis Wernwag used in his "Colossus" Bridge, 250 years later (see pp.76-7).**

Below and right **Following its destruction in the Second World War, the Ponte Santa Trinitá was rebuilt exactly as when first built, using the original materials wherever possible. These even included the pairs of statues from each end of the bridge representing the four seasons, which were recovered almost intact from the Arno.**

nitá, completed in 1567. Disdainful of the traditional Roman semicircle and the "barbarous" Gothic point, the architect, Bartolommeo Ammanati Batti-ferri da Settignano (1511-92), designed three near-elliptical arches with rises of only 1:7, but provided them with extremely shallow angles at their crowns to reduce the danger of collapse, and artfully concealed them behind decorative pendants. Despite its breathtaking beauty, the shape that resulted has since become known simply as "basket handled".

The masonry arch remained the most practical form for large permanent bridges throughout the Renaissance, but at the same time the speculations of some leading designers foreshadowed revolutions to come. In his *I Quattro Libri dell' Archittetura*, Andrea Palladio (1508-80) described and illustrated several types of timber truss bridge. In essence, a truss is a structural framework which can support weight. In the context of bridge design, it can be regarded as a beam (or indeed an arch) from which

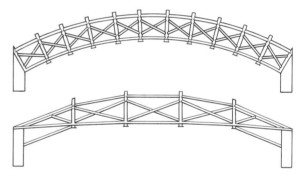

inessential material has been removed. Truss designs are innumerable, but all exploit the rigidity of the triangle and balance compressive and tensile forces to achieve the structural purpose.

Even more remarkable than Palladio's trusses are the bridge designs of Faustus Verantius, who published his *Machinae Novae* some time between 1595 and 1617. Like Palladio, Verantius explored the possibilities of trussed arches, but not Palladio's adaptation of his beam designs into arched form.

Verantius considered the possibilities of structures where the forces of tension and compression were self-contained. Traditionally, masonry arch bridges were massively buttressed to resist the downwards and outwards thrust of their self-weight, but Veran-

tius illustrated a bridge in which the outward thrust of the masonry arch was restrained by iron tie rods, themselves braced by further rods suspended from the arch. He then abandoned masonry completely, illustrating another bridge in which arch and deck were entirely cast in bronze.

Finally, Verantius came up with three suspension designs. The first was an elaborate version of the basket ropeway; the second was a temporary military bridge with ropes supporting a partially stiffened deck; and the third design – although subsequent generations were unaware of it – was an iron suspension bridge, its members composed of the eyebar tie rods used in his restrained arch structure.

Alongside these advances in new bridge forms, Renaissance scientists were evolving theories about the strengths of materials and the nature of forces acting upon them; and in his old age, Galileo began to develop the concepts and terminology which would transform structural design from an activity of empirical trial and error into one that was endowed with an increasingly precise theoretical and mathematical basis.

Below **Excavation taking place within a late 16th century French cofferdam, whose sides** are of a remarkably sophisticated dovetail design.

Below **Two of Verantius' theoretical designs: above, a tied arch of dovetailed and bolted timbers, reinforced with cross-bracing; in the background is what appears to be a precursor of Brunel's Saltash Bridge (see p.68); below is Verantius' celebrated iron suspension design.**

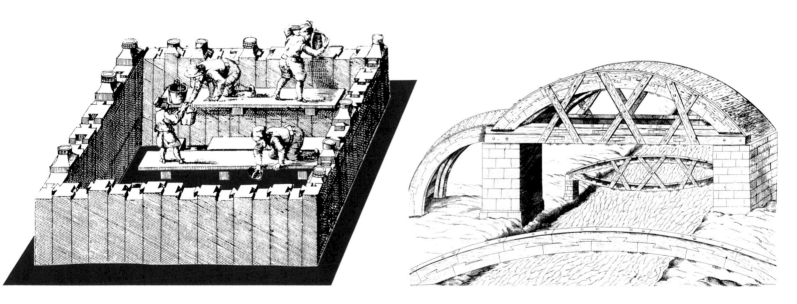

THE RIALTO BRIDGE, VENICE
ITALY

Most of the canals that thread like capillaries amongst the islands upon which Venice is built are narrow enough to be easily bridged, but the city's one great artery, the Grand Canal, was always the single major obstacle to both foot and vehicular traffic. A succession of timber bridges on the site of the present Rialto Bridge fell victim to inherent perishability until, following a fire in 1512, Fra Giovanni Giocondo, who designed the Pont Notre-Dame in Paris, suggested a permanent stone link. Designs were drawn up throughout most of the 16th century, some by the greatest figures of the Renaissance, including Palladio and Michelangelo (whose design is now lost). Nothing was decided until, in 1587, the Venetian Senate was moved to action at last, perhaps impelled by the perilous condition of the existing timber bridge. A new call for proposals brought

forth rival designs from the engineers Vincenzo Scamozzi (1552-1616) and Antonio da Ponte (1512-1597), already celebrated for restoring the Doge's Palace in 1577, after it had been damaged by fire.

Scamozzi followed Palladio in preferring a three-arch design, but da Ponte, who was then 75 years

Left and above **da Ponte's design was segmental, giving a rise of 6.4m (21ft): the deck was high enough to allow the passage beneath it of even the most imposing State barge, but sufficiently shallow and wide (22.9m/75ft) to accommodate two rows of 24 little shops, six above each haunch on either side of the bridge, with access walkways on the outsides behind the parapets, and a roadway down the centre.**

old, tackled the project with amazing vigour, and ambitiously proposed a single arch of 27m (88½ft) spanning the entire Canal from shore to shore.

Da Ponte's design was duly chosen, and the old man set to work. The two severest problems were the treacherously soft ground and the proximity of the foundations to the buildings on both sides of the canal. Da Ponte founded his bridge on 6,000 alder piles on each bank, driven within cofferdams in tightly packed groups, which became successively deeper the closer they approached the water. In addition, he cut off the piles in steps, reaching downwards and inwards towards the stream from the banks, and he surmounted them with broad brick platforms in corresponding shallow steps. On these, he placed radiating courses of abutment masonry, whose wedge-shaped blocks formed fanning bulwarks against the thrust from the arch and bridge superstructure.

Below da Ponte's foundations: tightly packed piles, with stepped platforms above. The bed joints of the masonry blocks, which formed the spandrels, continued on radiating lines from the focus of the arch, forming a beautiful correspondence between structural engineering and aesthetic ideals.

In June, 1588, when the foundations on the Rialto side had been completed and work was about to begin on duplicating them on the opposite bank, rumblings of doubt were heard about the safety of da Ponte's design. Encouraged and increased by intrigue, they eventually grew to such a pitch that the works were halted while investigations were mounted by the Senate. A commission of enquiry reviewed expert and not-so-expert evidence; and after intensive deliberation, it found in favour of the designer.

Work duly went ahead, and after another three-and-a-half years of intensive labour the Rialto Bridge was finally completed, in July, 1591. Almost immediately, it underwent the severest imaginable natural test, for in the same month Venice was rocked by an earthquake, which nonetheless left the bridge entirely unscathed – and it remains to this day, almost exactly as it was when it was built by Antonio da Ponte.

BEYOND EUROPE
ASIA

Below and left **Two Chinese bridges as different from each other in style and construction as structures could be: below is the stately curve of the Jade Belt stone bridge in the Summer Palace at Beijing. On the left, the zig-zag bridge leads to the Yu Yuan Bazaar in Shanghai. (Some claim that zig-zag bridges were designed to thwart devils, although in fact the zig-zag expresses a unity of opposites.)**

> "The bridge of jade arches,
> Like a tiger's back."
> *Li Tai Po*

While European engineers were refining old bridge forms and beginning to contemplate new ones, fascinating divergences and convergences were taking place on other continents. For example, some Chinese iron suspension bridges were using eyebar chains which were virtually identical to those in Verantius' unbuilt design (see p. 35).

Other, quite different types of bridge became important ingredients in the aesthetic, philosophical and spiritual unity which constituted the Chinese garden. Sometimes only simple arrays of stepping stones were used, but a more elaborate structure, and one peculiar to China, was the timber zig-zag, in which the ever-changing direction of the timber trestles symbolized the unity of opposites – the concept of *yin* and *yang*. Another constant feature of Chinese bridge design was the use of tall, slender arches, as on the famous Jade Belt stone bridge in the Summer Palace at Beijing, which also contains the Seventeen-arch Bridge, in which eight pairs rise successively from each side to the central and highest arch.

As early as the 7th century, the Chinese influence

Right **The simple but elegant Allahverdi Khan Bridge, with its 33 narrow pointed arches, was erected by Shah Abbas I of Persia, when he rebuilt Isfahan after making the ancient city his new capital in 1597.**

Below **The Kintai-Kyo Bridge at Iwakuni in Japan was originally built in the 17th century, although the regular rebuilding of each arch meant that the entire timber structure was renewed four times in each century. In 1953 the whole bridge was rebuilt in the same form, but its timbers were treated with modern, preservative chemicals.**

on garden and bridge design spread to Japan, where the tradition of allowing the walkway to follow the curve of the arch became equally well established. One particularly celebrated example is the five-arched Kintai-Kyo bridge over the Nishiki River in Southern Honshu. The timber bridge was built in 1673, and subsequently one arch of the bridge was rebuilt every five years.

Although virtually contemporary with the spare wooden arches of Kintai-Kyo, the splendidly arcaded Allahverdi Khan and Khaju bridges in Isfahan have an utterly different kind of beauty. Erected around c.1650, by Shah Abbas II, the shorter but more spacious and elaborate Khaju Bridge is a unique combination of a dam, a bridge and what is virtually a palace: two storeys of roads, shaded walkways, alcoves and galleries beneath tiled, pointed arches are carried on a deck 26m (85ft) broad and 141m (462½ft) long, itself built above a dam punctuated by arched brick sluices to control the flow of water, and edged by shallow flights of steps down to the River Zayandel. On both sides, at the centre and by each bank, there are six taller projecting pavilions, whose stately beauty completes what is as much a masterpiece of architecture as of engineering.

WESTMINSTER: A NEW BRIDGE FOR LONDON

ENGLAND

"London Bridge is falling down, falling down . . . ."

For more than 500 years, Old London Bridge was the capital's only crossing point over the River Thames, its narrow deck cluttered with houses, as much a hindrance to movement between the north and south sides of the City as the multiplicity of narrow, sluice-like channels between its many arches were to river traffic. Nevertheless, its uniqueness created such a flourishing concentration of trade that, at least among those who benefited, there was strong opposition to the construction of another bridge. The petitioners who set out in 1734 to campaign for the construction of a new bridge at Westminster had their work cut out to persuade Parliament to approve.

One of the most prominent of these petitioners was Nicholas Hawksmoor, the architect of some of London's finest churches. Hawksmoor was imbued with the classical ideals exemplified by the Pons Augustus at Rimini, but in his scheme for a bridge with nine semicircular arches, which was one of

Below **The central span of Charles Labelye's Westminster Bridge, showing how the centering would have been placed for erecting the voussoirs for the semicircular arch. The V-shaped cutwaters can be seen at the bases of the piers. The "turrets" above each pier enclosed pedestrian "refuges" on the footways.**

several schemes put forward during the 1730s, he also took account of the most recent European designs, such as the Pont Royal in Paris (1685) and the Blois Bridge (1724). In addition, and most importantly, his submission included the unprecedentedly scientific calculations of a young expatriate French engineer, Charles Labelye, on the effects of piers on river flow and water level on either side of a bridge, which caused much vigorous controversy even after Parliament had passed the necessary Enabling Act in 1736, the year of Hawksmoor's death. Eventually, in the following year, Labelye submitted his own design, for a 13-arch, all-masonry bridge.

In June, 1738, after twice being asked by the newly formed Westminster Bridge Commission to explain how he intended to found the piers, Labelye was asked to proceed but only with the foundations. His method was innovational in bridge construction. Adopting an idea which came in part from an earlier proposer, Batty Langley, he employed a re-usable timber caisson. After the location of each pier had

been dredged of gravel to make a flat surface, the caisson was floated into position above it. Then, while the first three courses of masonry were placed on its base, the caisson was continually flooded via sluices to sink it and then refloated by being pumped dry, so that it could regularly touch bottom and check that the surface remained flat. When the first three courses were loaded, the caisson was sunk into position and more masonry was added until the pier stood well clear of low water. Finally, the sides of the caisson, which were dovetailed and wedged into the base, were knocked clear of it, floating to the surface to be used again, while the base remained as a foundation for the pier.

While Labelye was working on the foundations, a timber superstructure was commissioned from another contractor, but the order was later cancelled, and in 1740 work began on Labelye's stone design. The arches were complete by 1747, but then one pier began to settle more than the rest and it went on doing so. The arches bearing on it were taken down, new piling was driven round its edge to stabilize it, and the arches were rebuilt to be thinner and with covered-in voids. The repair was successful, and Westminster Bridge opened on 18 November, 1750.

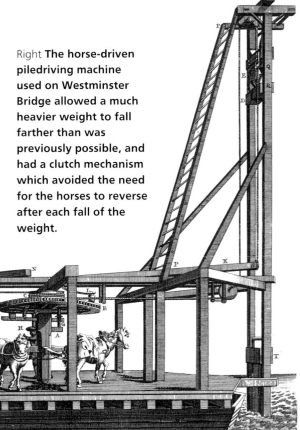

Below **With Labelye's stately arches Westminster had a Thames crossing to match, even surpass, that of the City of London. In this contemporary painting, only the bulk of Westminster Abbey challenges the impact of the new bridge.**

Right **The horse-driven piledriving machine used on Westminster Bridge allowed a much heavier weight to fall farther than was previously possible, and had a clutch mechanism which avoided the need for the horses to reverse after each fall of the weight.**

PERRONET AND THE PONT DE NEUILLY, PARIS

FRANCE

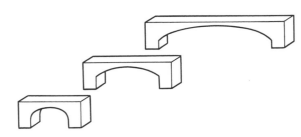

Under the influence of King Louis XIV, a nationwide road system became a priority in France, leading to the establishment in 1716 of the first professional engineering body, the *Corps des Ponts et Chaussées*; and in 1747 the need to educate the engineers led to the foundation of the world's first school of engineering, the *École des Ponts et Chaussées*, under the direction of the able and experienced bridge-builder, Jean-Rodolphe Perronet (1708-94).

In 1763 Perronet supervised the construction of a bridge at Mantes on which the ratio between the spans and the width of the piers was an advanced 5:1. The designer, M. Hupeaux, had followed the example of the leading architect Jules-Hardouin Mansart, who had introduced this reduced ratio on the Pont Royal in Paris, thereby demonstrating that the combination of a Roman semicircular shape and a span/pier ratio of 3:1 had been an unnecessarily conservative recipe for free-standing stable arches. While completing Hupeaux's bridge, Perronet noticed that a pier supporting a completed arch was leaning a little in the direction of its yet-to-be-completed neighbour. It was a clear indication that horizontal thrust was present, but the fact that such bridges stood satisfactorily when completed had to mean that these horizontal thrusts were carried along the whole extent of the bridge, and into the abutments at the banks.

Then Perronet made a simple but profound leap in comprehension. Why not turn this phenomenon to both practical and aesthetic advantage by embracing it – make the arches flatter, to carry yet more lateral force into the abutments, and build much more slender piers, thereby simultaneously giving the bridge a more elegant profile and reducing the obstruction to both the current and the river traffic?

The great bridge on which he triumphantly proved

Left **Growing span-pier ratios. Front to back: the Roman 3:1; the 5:1 at Mansart's Pont Royal, Paris and Perronet's own achievement of virtually double the ratio, at 9:1.** Above **The five completed arches of the Pont de Neuilly across the Seine, with the *cornes des vaches* visible at the bases of the piers. Some have hailed the Pont de Neuilly as the most graceful and beautiful stone bridge ever built.**

his theory was the Pont de Neuilly, which he began to build across the Seine to the north of Paris in 1771. It required five spans of 36.6m (120ft) each, in which Perronet used a development of the *cornes des vache* device. Flattened elliptical curves to the arches at the bridge's centreline, springing from near water-level, were brought out and up on each side, shaved off by the *cornes des vaches* to become very shallow segmental arch profiles on the facades, which sprang from points over 5m (16ft) higher than the central curves. This sophisticated and elegant effect was enhanced by the great revolutionary feature of the Pont de Neuilly, the ultra-slender piers of only 4m (13ft), giving an unprecedented ratio of 9.3:1.

The necessary corollary of the design was, of course, the erection of wooden centering for the entire bridge, and construction of all the arches simultaneously. This took place during a single year, and on 22 September, 1772, Perronet, obviously something of a showman, turned the striking of the

Above and Left **Two engravings of the Pont de Neuilly under construction; above Perronet engaged water power to operate his machinery. Paddle wheels powered by the river current operated bucket wheel "trough pumps" (labelled "C") to drain the cofferdams; the bridge piers were subsequently founded on timber piles within these cofferdams. Left The centering for three of the arches (labelled "C") simultaneously under construction, springing from near the bases of the partially completed bridge piers.**

centering along the entire length of the bridge into a piece of spectacular theatre, by so organizing it that the whole lot tumbled into the water in only a few minutes. Cheers arose from the assembled multitude and King Louis XV, who had expressed a particular desire to be present, rode across the bridge in his carriage. Despite all Perronet's care with the foundations to the Pont de Neuilly, the bridge still settled slightly when the centering was removed. Nevertheless, it stood for nearly two centuries, until it was demolished in 1956 in an act of particular historical vandalism.

Perronet's next bridge, the Pont Sainte-Maxence, though smaller, was even more daring. The ratio of span to rise was more than 11:1, and this time the piers, instead of being solid across their width, were each divided into two 2.7m (8ft 8in) columns linked by a small arch, which gave an even lighter effect to the whole. His final masterpiece was the Pont de la Concorde in Paris. By then in his 80s, Perronet nevertheless supervised the construction of the bridge while the storm of the French Revolution swirled about him. Indeed, it even contrived to provide him with materials, as the Pont de la Concorde incorporates masonry from the demolished Bastille.

THE IRONBRIDGE, COALBROOKDALE

ENGLAND

Although Charles Labelye's methods of founding the piers of Westminster Bridge incorporated new technology, the shapes of his arches were Roman. When, therefore, in 1759, Robert Mylne (1734-1811) presented a design with elliptical arches for the next bridge in London (at Blackfriars) to be approved, conservative heads shook; and their doubts were reinforced, with more eloquence than engineering knowledge, by no less a man than Dr Johnson. As built, Mylne's nine arches for Blackfriars Bridge still had modest ellipses, and for the foundations he used an improved version of Labelye's floating caisson, but in general his masonry arch designs seem tame beside the brilliance and daring of Perronet.

Somewhat the same can be said of the other leading English bridge-builder of the mid-late 18th century, John Smeaton (1724-94), although, judging by his writings and his unbuilt designs, Smeaton was neither timid nor backward-looking. Ten years before Perronet struck the centering of the Pont de Neuilly, Smeaton wrote, "I look upon it, that no limit to the span of arches, in proportion to their rise, has yet been found; since the widest and flattest arches that have been attempted, upon right principles, have succeeded as well as the narrowest and highest; provided the abutments are good, and the stone and cement . . . are of a firm texture."

Among Mylne's designs there is one small but striking exception to his sequence of solid, worthy masonry arches. In 1774 he sketched a design, which remained unbuilt, for a small bridge of two joined arches at Inveraray. It was very light and very shallow, and it was made of *iron*. At about the same time, Thomas Pritchard of Salisbury produced a sequence of three designs: a timber trussed arch; a masonry bridge on a cast-iron centre; and a fully cast-iron arch between brick abutments. The third, a "design for a cast iron bridge between Madeley and Broseley" was the first version of what is commonly regarded as one of the seminal bridges, the Ironbridge at Coalbrookdale (1777-79) – the first cast-iron arch span and now the focus of a national museum.

Pritchard's original design was considerably modified – among other·changes, the roadway became wider, and the arch higher but reduced in span – but once the casting of the members began at the Coalbrookdale company, an iron bridge was erected within only three months.

Although cast iron is stronger in tension than masonry, and also most timber, its real advantage is its greatly superior compressive strength. This is reflected in the slender ribs and connectors of the Iron-

The Industrial Revolution grew from Coalbrookdale in Shropshire, where blast furnaces to smelt iron ore had begun to operate in the 17th century. As the industry developed on both sides of the River Severn, a bridge link became increasingly necessary, and as the gorge at this point is fairly narrow and steep, it had to be a single high arch. Naturally, the local material, iron, was chosen, although construction in iron was still in its infancy – the mortise joints and the pegged dovetails which respectively link the main arches to uprights (above right) **and to radial members** (below right), recall timber jointing techniques. Each of the five main semicircular ribs was cast in two halves, each weighing nearly six tons. In the decades following completion, pressure from the abutments pushed the two halves together, raising the centre of the roadway by a few feet (facing page).

bridge – which is structurally an arch in compression. Indeed, its builders did not yet really comprehend the potential of the material: it contains a large amount of redundant cast iron, and the jointing echoes timber practices in its dovetails and mortises. The Ironbridge is, in fact, significant almost in spite of itself. Within a few years, iron bridges were to be designed and built that were truly representative of the Industrial Revolution, developing bridge technology *per se* and exploiting the possibilities of the "new" material to the full.

THE IRONBRIDGE	FACTS
bridge completed	fall 1779
span of arch	30.5m/100ft
total weight of iron	378½ tons
principal builder	Abraham Darby

part two

At the start of the Industrial Revolution, the forms and constructional details of the first iron bridges still humbly followed stone and timber precedents from past centuries, but by little more than a hundred years later, colossal steel structures in bold new shapes, fully exploiting the strengths of the material, had been thrown across hitherto unbridgeable gulfs throughout the Western world.

Perhaps more than any other structure, the Forth Rail Bridge symbolizes the mighty strides made by bridge engineers in little over a century of industrialization.

5
THE INDUSTRIAL REVOLUTION

"This extraordinary metal, the soul of every manufacture, and the mainspring perhaps, of civilised society."

Samuel Smiles, *Invention and Industry*

The iron used for the Ironbridge was already the product of a century's local refinement in smelting methods. What were new were the scale and purpose of the casting — not so much a revolution in bridge design as an an evolution of applied technology.

The next big step forward in the use of iron for bridges stemmed from a man now remembered as a (political) revolutionary, Thomas Paine. After taking part in the United States' fight for independence, Paine proposed the construction of long iron spans, to avoid the necessity of founding arch piers in ice-clogged rivers. In 1787, he went to London to raise support for his proposals. In 1789 a 27.4m (90ft) trial rib, which he designed, was fabricated and load-tested. In 1790, a complete 36 ton bridge in wrought- and cast iron segments was brought to London and displayed on Paddington Green for over a year, but by the time others began to propose a full-scale structure based on his patent, Paine's mind had moved on.

In the early 1790s, a rich landowner called Rowland Burdon took over a project to build a bridge across the Wear at Sunderland, which in his opinion required a single, very long span. With his friend and advisor, the architect Sir John Soane, Burdon had become familiar with the covered timber truss bridges erected by the Swiss Grubenmann brothers, which were the climax of a wave of interest in timber truss bridges in the first half of the 18th century generated predominantly by Palladio's designs. These bridges proved the practicality of spans much longer than those that were then thought feasible in masonry, and the examples of the bridge at Coalbrookdale and Paine's prototype on Paddington Green demonstrated a new way forward with iron. Burdon engaged a local builder, Thomas Wilson, under whose supervision six huge ribs were fabricated and were erected over the Wear in only ten days in 1796.

The first decades of the 19th century were dominated by two Scottish engineers, both of whom responded to the challenge of iron. One was John Rennie (1761-1821), a farmer's son from East Lothian, who graduated from Edinburgh University in 1783 and went to work for the Birmingham firm of Boulton & Watt. In 1790 Rennie went to London to supervise the machinery at their new steam-powered Albion Mills, but he soon became involved in other engineering projects, including two proposals for iron bridges of roughly the same span as the Ironbridge at Coalbrookdale, which were to be prefabricated in sections and shipped to the island of Nevis in the West Indies. For several years after that his bridge-building activities were confined to masonry structures, but his contribution to the development of iron spans was far from over.

The second Scot, Thomas Telford (1757-1834), although entirely self-taught, became perhaps the most versatile engineer of his generation. After working in Edinburgh, London and Portsmouth, he was surveyor of public works for Shropshire from 1787. His first significant bridge, completed in 1792, was at Montford on the Severn, and consisted of three elliptical spans in sandstone. In 1795, when

Above **Thomas Paine.**
Right **Ulrich Grubenmann's covered timber truss bridge at Schaffhausen, 1757.**

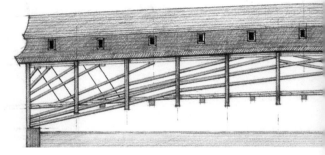

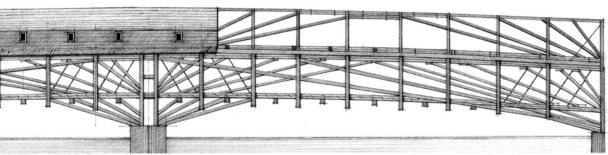

Above **Scaffolding supporting the main ribs of the Wear Bridge. Although the bridge was well over twice the span of the Ironbridge, and shallower, with a wider roadway, it used only 260 tons of iron.**

Telford's Bonar Bridge was triumphantly successful. In the remaining years of his long career, he designed further successful iron arches, the most beautifully sited of which was undoubtedly the Craigellachie Bridge (1815) (left), **which crosses the River Spey in northern Scotland. On first seeing it, the poet Southey wrote, "As I went along the road by the side of the water I could see no bridge; at last I came in sight of something like a spider's web in the air... and oh, it is the finest thing that ever was made by God or man!"**

severe floods brought down the old Buildwas Bridge on the same river, he seized the opportunity to design a replacement in iron. This was quite different from both the Ironbridge and the Sunderland Bridge. It had a narrow 5.5m (18ft) roadway supported by three ribs of shallow curvature, each cast in three sections, spanning 39.6m (130ft), with a rise of only 5.2m (17ft). The ribs were intersected by two much deeper arches on the outside, springing lower down and rising at the centre nearly to the tops of the guard-rails.

By the turn of the century, the prospects for cast-iron arch bridges seemed boundless, but the next few years saw the collapse of some, and major problems with others, including the Sunderland Bridge. To Telford goes the credit for revitalizing the form. In 1803 he had become engineer to the Commissioners for Roads and Bridges in the Highlands of Scotland, and in 1811, after many stone bridges, he designed a crossing for a narrow point on the Dornoch Firth. The Bonar Bridge was an innovatory 45.7m (150ft) arch, whose four principal girder ribs were each pre-cast in five large lattice sections, rather than assembled from many individual beams.

THE ROMANS REBORN: PONTCYSYLLTE

WALES

A canal-builder, faced with the need to carry a waterway across a valley, has two choices: a series of descending and ascending locks, or an aqueduct. In 1759, the Duke of Bridgewater decided on the latter for Britain's first wholly artificial canal across the Irwell Valley in Lancashire. The resulting Barton Aqueduct, designed by James Brindley, became a model for many similar structures, as Britain's canal network grew during the remainder of the century. The Bridgewater Canal was carried over the river itself on three low segmental arches in a trough made of waterproof puddle clay – a worked mix of clay, sand and water.

Brindley's successor as the country's leading canal engineer was William Jessop; and it was his recommendations that gave both Rennie and Telford their crucial canal engineering appointments. Of Rennie's many aqueducts, the one built in 1798 to carry the Lancaster Canal over the Lune on five, massive, 21.3m (70ft), semi-circular arches remains not only his Classical masterpiece, but also the largest masonry aqueduct in Britain.

By this time, Thomas Telford had designed and built the first all-iron aqueduct at Longdon-on-Tern on the Shrewsbury Canal. However, Telford's first great achievement – and the one that elevated him to

Above **Thomas Telford (1757-1834).**

Left **The massive Lune Aqueduct demonstrates the enduring popularity of Classical forms in the Industrial Revolution.**

Right **The Longdon-on-Tern Aqueduct, 1795-96. Although small in scale, with four spans totalling less than 60m (200ft), its three slender cast-iron supports of uprights and diagonals plus rectangular trough are thoroughly conceived in terms of the material and not as an *ersatz* masonry structure.**

Facing page **Pontcysyllte remains unique in scale and magnificence – and thanks to the quality of the iron, has withstood the test of time.**

the level of the nameless and timeless Roman masters of the Segovia Aqueduct and the Pont du Gard – was Pontcysyllte, completed in 1805, to carry the Ellesmere Canal over the Dee Valley near Llangollen. Everything about it speaks in superlatives.

Approached by a long embankment, the aqueduct itself extends for 307m (1,007ft) at a maximum height of 38.7m (127ft) above the river. Nineteen cast-iron arches, each spanning 13.7m (45ft), are carried by stone piers, constructed solid to a height of 21.3m (70ft) and then hollow above, which not only reduces weight but also provides access for inspection and maintenance. Four ribs, cast in three sections each, comprise the arches, supporting trough units 3.61m (11ft 10in) wide. Both the arches and the side walls of the trough have a similar, but more pronounced, patterning to that at Longdon-on-Tern.

Water fills the full width and height of the trough, to within inches of the edge. Although about a third of the trough is covered over by the iron plates of the tow-path, which narrows the navigable width, the extra room underneath the plates allows the water to slip past boats with less drag resistance than would be possible if the tow-path were outside the channel, which was the case at Longdon-on-Tern and on at least one smaller, subsequent iron aqueduct.

SPANNING THE THAMES IN MASONRY: LONDON
ENGLAND

Old London Bridge in medieval times, Westminster in the 1740s, Blackfriars by 1769 . . . it had taken nearly six hundred years for London to acquire three crossings of the Thames, but within only three years (1816-19), the number doubled. The first of the new bridges to be completed was Vauxhall at Lambeth. This had nine relatively modest iron arches, each spanning 23.8m (78ft) and was designed by James Walker, who had taken the project over when a masonry design by John Rennie was aborted due to cost. Nevertheless, Rennie became directly responsible for the next two Thames bridges, and posthumously for a third.

By the 1810s, his experience in masonry bridge work was extensive and his opinions were strong as well. When asked to comment on a design based on Perronet's Pont de Neuilly, prepared for the Strand Bridge Company, he and William Jessop disapproved of the *cornes de vache* and wrote, "Our opinion therefore is, that the bridge over the Thames should either be a plain ellipsis without the slanting off in the haunches so as to deceive the eye . . . or it should be the flat segment of a circle . . ."

The "bridge over the Thames" became Waterloo Bridge, and the nine granite arches, designed by Re-

Above **One of the nine granite-clad arches of the Waterloo Bridge. Each arch spanned 36.6m (120ft) between 6.1m (20ft) thick piers. The Venetian sculptor Canova, who visited London near the time of the Bridge's opening, called it "the noblest bridge in the world, worth a visit from the remotest corners of the earth", and it became perhaps London's best-loved bridge. Much protest accompanied its demolition in the 1930s (as a result of subsiding foundations), and the replacement of it by a new Waterloo Bridge.**

nnie after the Strand Bridge Company gave him the job in 1810, were indeed "plain ellipses". His spandrel walls reflected another criticism of Perronet: instead of the French designer's plain (and plane) surfaces, Rennie incorporated three-quarter columns over the piers, the single square pilasters of his Lune Aqueduct now transmogrified via many intervening projects into pairs of Doric supports to projecting refuges on the parapets, like a cortège of severe temple entrances. Rennie's design did accord with that of Perronet in one respect – he provided a virtually flat roadway from the South Bank to the Strand.

Rennie had intended to found the Waterloo piers on floated caissons, as Labelye and Mylne had done at Westminster and Blackfriars, but after experimenting with cofferdams at the shallow south end, he decided to use these throughout. Two steam engines were used to operate pumps, and one of them may also have been used to drive some of the piles, although most of them were put in place by manually-operated, geared machinery. In constructing the arches, Rennie introduced prefabricated ribs for the centering, floating them into place between the piers on barges, and as each arch was completed, removing and re-erecting them for the next one.

By the time Waterloo Bridge was opened, in 1817, work was already well advanced on John Rennie's other great Thames project, Southwark Bridge. The site chosen was a narrow and deep point on the River, and one which could be reached by ships should Old London Bridge upstream ever be demolished. The difficulty of founding, and the need to avoid impedance, led to the design of a massive three-span cast iron structure, with a central span which, at 73.2m (240ft), was the largest and longest yet erected and had a rise of 7.3m (24ft). Again Ren-

Above Top **Southwark Bridge had Britain's largest cast-iron arches: the centre span alone weighed 1300 tons.**

Bottom **Waterloo Bridge in its Regency prime.**

nie used cofferdams to found the piers, bounding them with double rows of timber piles which were even more massive than those used at Waterloo; and steam engines, again more powerful than before, pumped out the water from the cofferdams.

Huge granite blocks, weighing up to 25 tons, were quarried and brought down from Scotland for the piers. The ironwork was comparably massive – and by common consent considerably over-designed. The eight ribs of each arch were cast solid, in 4m (13ft) lengths – in effect vast voussoir units.

REPLACING LONDON BRIDGE
ENGLAND

Old London Bridge consisted of 19 pointed arches, spanning between 4.6m (15ft) and 10.5m (34ft), none exactly the same, and carried on a similar assortment of piers. The Thames at this point is relatively narrow and tidal in both directions, which made construction extremely difficult and hazardous. The bridge was built an arch at a time, sprung from piers founded on boat-shaped cutwaters (starlings) so massive that far more of the width of the river was filled in than was spanned by the arches. Fierce currents were forced through the narrow openings, and constant repairs for the next six centuries had been the result.

In 1746, while Westminster Bridge was under construction, Charles Labelye was commissioned to conduct a survey of Old London Bridge. He proposed to remove the starlings and enclose the piers with stonework, but not to widen the bridge or get rid of the houses which now lined its sides. The City authorities appointed a committee to report further; and this time the main proposals embraced precisely the two elements side-stepped by Labelye. The plans, drawn up by Sir Robert Taylor and George Dance the elder, included a further important feature – the replacement of the two centre arches and their supporting pier and starling with a single navigation

Below **One of the great unbuilt designs of engineering history was Thomas Telford's amazing 1800 proposal for a single 183m (600ft) cast-iron arch to replace Old London Bridge.**

span. This, plus removal of the houses, partial widening and new spandrel walls, was completed in 1763 – and before the year was over, John Smeaton had to be called in to design emergency remedial measures to stop scour from the newly enlarged central channel demolishing the rest of the bridge.

A flurry of proposals in 1799-1800 for a replacement culminated in a single-span design by Thomas Telford. Though this was not used, Old London Bridge finally ran out of credit only 20 years later: its piers and starlings were damaged by boats, and needed constant expensive repairs; its roadway, even

Left **London Bridge as it is today:** a triple-arched, prestressed-concrete bridge; elegant and efficient, but never to be loved by the capital's inhabitants as were its two predecessors.

Below **Rennie's New London Bridge of 1831** reflected the ever-growing technical confidence of the time. His five arches, 9m (30ft) above high water, spanned 39.6m, 42.7m, 46.3m, 42.7m and 39.6m (130ft, 140ft, 152ft, 140ft, 130ft).

denuded of houses, was inadequate for the ever-growing commercial traffic of the City. The Corporation grasped the nettle and consulted with eminent engineers and architects. Their plan, composed of repairing, strengthening and replacing eight of the arches with four, was referred to John Rennie, who carried out a typically thorough survey of the old Bridge and the state of the river up and down stream.

Rennie's conclusion was unequivocal: demolition and replacement with a new London Bridge. The authorities deliberated, discussed, agreed, and applied for an enabling Act of Parliament. Rennie prepared a detailed design, consisting of five semi-eliptical arches, but he died soon afterwards, on 4 October, 1821, just after the Act had been passed. A design competition was held, but despite the submission of over 30 entries, Rennie's design was adopted; and its building was placed in the hands of his son, John, who was later knighted (although his brother, George, was the more considerable engineer). New London Bridge was opened in 1831 (Charles Dickens a fascinated observer); and it survived until its replacement in 1972, when it was dismantled and re-erected in Lake Havasu City, Arizona.

Below **A very early engraving of Old London Bridge** at its most familiar, crowded with houses, its mismatched cutwaters greatly impeding the River.

Right **Old London Bridge after 1763** – denuded of the familiar jumble of houses, built and rebuilt over the centuries, and with a new navigation span where two arches had formerly been.

THE FIRST RIGID SUSPENSION BRIDGES

It was an American, James Finley, who first identified the major components of the modern suspension bridge: a system comprising main cables and vertical suspenders from which was hung, most crucially, a level deck braced by trusses. A judge and a justice of the peace in Fayette County, Pennsylvania, Finley built his first bridge, a 21m (69ft) span across Jacob's Creek, in c.1800. In 1808 he patented a suspension system, and in 1810 he published a "description of the Patent Chain Bridge" in a New York journal, *The Port Folio*.

By the time of his death in 1828, Finley had completed at least 13 bridges. His second, in 1807, spanned 39m (128ft) over the Potomac, and two years later he achieved his first multi-span chain bridge at Schuylkill Falls, near Philadelphia. By the standards that were soon to be achieved elsewhere, Finley's bridges were small, crude and vulnerable. The Schuylkill Falls one collapsed in 1811, as did a replacement in 1816. In the same year and on the

Above **The first suspension bridge to have a horizontal deck made rigid by the incorporation of a truss. Its designer, Judge James Finley from Pennsylvania, published his Patent Chain Bridge in *The Port Folio* in 1810. The deck is supported by a chain on either side, but the fact that the catenary drops below deck level at the centre shows that this part of the deck rested on the chains, rather than being suspended from them. The beginning of the second span is visible on the left-hand side.**

Left **Anchorage for the chain at one extremity of Claude Navier's ill-fated Pont des Invalides, a suspension bridge across the Seine in Paris that had to be demolished before it was completed, in 1826, due to movement in one of the abutments.**

Right **A semicircular stone archway leads onto the 5.5m (18ft) wide roadway of the Union Bridge across the Tweed.**

same site, however, others constructed a tremendously important harbinger of achievements to come – the first wire-cable suspension bridge, albeit a temporary, unstiffened one for pedestrians. It had a span (or spans) covering over 120m (393¾ft), but it was only 0.5m (1½ft) wide.

The most substantial Finley-type bridge was the single 74m (243ft) span Merrimac Bridge, built under license by John Templeman in 1810. Ten chains, suspended between the 11.3m (37ft) high stone abutments, supported two 4m (13ft) roadways on hangers. As the *Newbury Port Herald* of 14 December, 1810, affirmed, "Horses with carriages may pass upon a full trot with very little perceptible motion of the Bridge". The Merrimac Bridge collapsed under snow in 1827, but it was rebuilt and survived until 1913, when it was replaced with unwonted zeal by a near-replica in concrete and steel and the last original design of "The Father of the Modern Suspension Bridge" was lost.

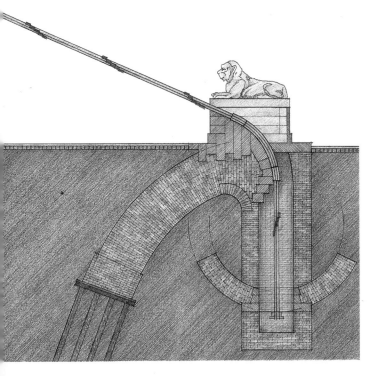

Above and right **The Union Bridge across the Tweed, linking Scotland and England, was built in 1820 by Captain (later Sir Samuel) Brown, whose interest in ships' rigging led him to develop and patent the type of circular eyebar suspension chain used here: the deck is** supported by vertical rods suspended from three pairs of wrought-iron chains on each side. Although the Union Bridge is the earliest major suspension bridge, the chains have survived intact – although they are now assisted by modern steel cables.

THE MENAI STRAIT BRIDGE

WALES

"I heard him then, for I had just
Completed my design
To keep the Menai Bridge from rust
By boiling it in wine."
The White Knight in
Alice Through the Looking-Glass.

The concept behind James Finley's patent system (see p.56) was first taken up and developed in Britain. In 1814 and 1818 respectively, visionary suspension schemes were proposed by Thomas Telford and John Anderson, for the Mersey at Runcorn and the Firth of Forth. However, the first to translate vision into substantial structure was Captain Samuel Brown R. N. (see pp.56-7), who was later knighted.

Brown and Telford both experimented with various types of wrought-iron linkages. Indeed, Telford's Runcorn proposals, which postulated both bar and wire cable designs, were made only after extensive tensile strength tests. Due to their cost, these came to nothing, but Telford's next suspension design did come to fruition.

A bridge was urgently needed across the Menai Strait between the Welsh mainland and the island of Anglesey, to carry traffic bound for Ireland to the port of Holyhead. Since 1776 there had been various proposals, including ones for cast-iron arches from Rennie and Telford; Telford planned to construct the centering by suspending it from above, but in 1818 he produced a suspension design of unprecedented scale and magnificence. The main span

Right **With its seven imposing masonry arches, four on the Anglesey side, three on the Welsh, the Menai Bridge is both an architectural and engineering masterpiece, though the robust truss deck and the replacement of the iron chains with four thick steel chains diminishes the original contrast between arch and suspension elements. Despite its overall scale, rigidity, and structural mastery, it was never the world's longest unsupported span: the Taoguan (Peach Pass) bridge in Szechuan, China, built in 1776, spanned 200m (656ft).**

alone was to be much longer than Brown's Union Bridge; and whereas that had been hung from squat towers founded on the river banks, the Menai was to have towering masonry approach viaducts supporting the roadway beneath long spans of anchoring chains.

Construction of the stone superstructure began in 1820 and continued until 1824. As with the Pontcysyllte, Telford left large voids inside the piers, preferring interior cross-walls to rubble fill for stability. For the next stage, the erection of the side-spans, tunnels were excavated on each bank and joined underground to form caverns, and then massive cast-iron frames were placed within them to make secure anchorages for the chains. By the spring of 1825 the side-spans were complete from their anchorages to the tops of the towers. All that remained was the erection of the main span.

One chain had been left hanging down the tower on the Welsh side, and now the remainder of the first centre span was loaded on to a long raft. On 26 April, 1825, the raft was manoeuvred out and moored between the towers. One end of the chain was bolted to the hanging piece, and the other fastened to ropes passing up and over blocks on the top of the Anglesey tower to capstans on the shore beyond. Thousands of spectators waited and watched on boats and on the shore. At a shouted signal to "go along!", a band started up and 150 labourers strained at the capstans. The chain rose into the air, and in only 95 minutes was in position,

Right **After construction began on the Menai Bridge, Telford was commissioned to build a similar span across the Conway River, a few miles away. Completed in 1826, it has a 99.7m (327ft) deck suspended from 10 (2 x 5 grouped vertically) wrought iron chains between towers quaintly castellated to match the Castle.**

bolted to the side span atop the Anglesey tower. By July 9, the remaining 15 chains had been put in place without mishap. The square-section suspension eye-bars were fixed, the roadway was constructed, and on 30 January, 1826, the London to Holyhead mail-coach passed over the Menai Bridge for the first time.

Telford's planning proved triumphantly right in every important particular except one – the wind-resistance of the deck. Undulations were experienced and remedial measures taken even before the bridge opened, but several times in 1826 gales caused more problems. Transverse chain-bracing was added, but in 1839 the most severe storm yet broke many of the

hangers and wrecked the deck. It had to be com-pletely rebuilt, as it was again in 1892, and again be-tween 1938 and 1941, this time with the substitution of steel chains for iron.

THE MENAI STRAIT BRIDGE	FACTS
total length	521m/1710ft
suspended span	176m/579ft
total height of towers	46.6m/153ft
elevation of deck above Strait	30.5m/100ft
number of chains	4 x 4, grouped vertically
length of individual eyebars	2.9m (9ft 6in)
uniform span of arches	16m (52ft 6in)

THE GRAND PONT SUSPENDU, FRIBOURG
SWITZERLAND

Thomas Telford was the first to consider wrought-iron wire cables as an alternative to chains, but he never used them on a suspension bridge, and although a number of small wire-suspended bridges were built in Britain and the United States during the 1810s, the next real development in the design of suspension bridges took place in Europe. The French engineer Claude Navier published a highly influential analytical treatise early in the 19th century; and the Swiss Guillaume Dufour and the five French Seguin brothers experimented with drawing and

Below **Joseph Chaley's world-beating bridge in its hey-day. Never has a longer single span been constructed in Switzerland.**

linking wires during the 1820s. The Seguins' first small test bridge was erected at Annonay in 1822; in 1823, the eldest brother, Marc, built the world's first permanent wire-suspended bridge in Geneva; and over the next 20 years the Seguins, Dufour and others built several hundred throughout Europe.

In France, drawn iron was easier and cheaper to manufacture than the chains favoured in Britain. It needed no bolted connections and it used much less metal, as it possessed roughly double the tensile strength of equivalent forged bars. However, wire

was more prone to rust, and while developing methods of splicing wires together and linking cables, European engineers also investigated ways of preventing corrosion, including dipping the wires in oil, painting them and cladding them in sheet metal.

It was Joseph Chaley, an obscure engineer who had previously worked with the third Seguin brother, Jules, who designed the next record-breaking suspension bridge – a span which, at 273m (896ft), was more than half as long again as the Menai Bridge. This was the "Grand Pont Suspendu", built over the precipitous Sarine Valley at Fribourg in Switzerland, in 1834. A bridge was essential here in order to ensure that the lucrative traffic between Berne and Lausanne continued to flow through the

Elevation (above) and **plan** (below) **of the drum around which each strand of cable for the Grand Pont Suspendu was spooled. Each cable was doubled, then wound onto the drum, leaving the two ends free. Each end was then hauled in opposite directions, outwards and upwards towards the tops of the towers on the gorge sides, the drum revolving as the strand unwound. The ends were then fixed to the backspan cables.**

town, and from 1824 onwards a number of proposals were submitted, including several masonry-arch and timber designs and an "underspanned" suspension bridge by Dufour. All were rejected, however, either because they were too expensive or because they were unlikely to be durable enough, and then, in 1830, Chaley arrived with alternative proposals. The first was for twin wire-supported spans with a central masonry pier, and the second was for an unprecedentedly long single span. As the very high pier on his first proposal made it much more expensive, Chaley was very quickly given the "go-ahead" for the long span.

On previous wire suspension bridges the cables had been prefabricated in one piece: the wires were laid parallel and bound round by other wire and were then hauled into place from the ground beneath by ropes. For the Fribourg bridge, however, the four cables designed by Chaley were far too long and heavy to be made as single units. Each was made from over 1,000 wires grouped in 20 strands. When they were all in position, the strands were bound together to make a single cable. On the tops of the towers were hollow, cylindrical, cast-iron "saddles", over which the cables were pulled and then attached to the "backspan" cables, which were themselves anchored into the rock on each side of the bridge by cast-iron and rock wedges.

The Grand Pont was an extraordinary achievement. Its railings were trussed, which seems to have stiffened the deck enough to avoid disconcerting oscillations, and the distance between the cables on either side was greater than the width of the deck, with the result that the suspenders drew them in and further braced the whole structure. If the bridge had an "Achilles' heel", it was the anchored sections of the cables, which, when checked only 16 years later, were found to have corroded greatly inside their wedges. Extra cables were then added in such a way that the anchorages could be checked regularly, and after that the Grand Pont survived until 1923, when it was replaced by a multi-span concrete bridge.

Only in one respect was the bridge the "last of the old" rather than the "first of the new". Another French engineer, Henri Vicat, devised a way in which wire could be continuously unspooled in the air, and Chaley subsequently developed it on other bridges, but it was on the opposite side of the Atlantic that this new technique was to come into its own.

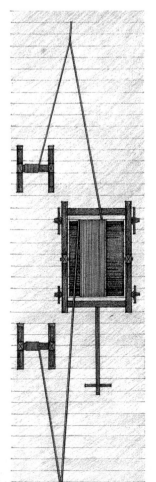

6

THE VICTORIAN ACHIEVEMENT

The first railway bridge, the Causey Arch at Tanfield in County Durham, had already been in existence for nearly a century when George Stephenson (1781-1848) began to experiment with steam-driven locomotives. A near-semicircle of masonry, spanning a respectable 31.3m (103ft), the bridge was built in the 1720s to carry a double wooden railway for coal wagons – and although it is still standing, the fear of its collapse is said to have been the cause of its designer's suicide. But the wagons were, of course, horse-drawn. In the early 19th century, bridges were built only to carry pedestrians, horses, horse-drawn vehicles and man-made waterways. None was constructed to carry the supreme embodiment in motion of that Revolution – the steam railway.

After 1815, however, George Stephenson, "the Father of Railways", turned the ponderous, halting "steam boiler on wheels" of the 1810s into a machine capable of pulling a load many times the capacity of the finest team of horses. In 1825 he completed the world's first locomotive-drawn goods and

Above **The High Level Bridge, Newcastle-upon-Tyne (1849). The railway runs along the upper deck, which is supported by six 38.1m (125ft) cast-iron bowstring arches; beneath, a roadway is carried on wrought-iron suspension rods hanging from the arches.**
Right **In a bowstring arch (right) the outward thrusts are restrained by a horizontal tie linking the ends.**

passenger line from Stockton to Darlington. Five years later, his Liverpool to Manchester Railway was opened. From then on, "Railway Mania" spread through the country, reaching its peak in 1845; and it has been estimated that, during the most intensive years of the "Mania", the number of bridges in Britain virtually doubled, from 30,000 to 60,000.

More than any others, three men drove the vast

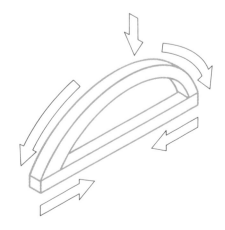

on to supervise the construction of the eastern section of the Liverpool-Manchester Railway, before being given overall responsibility for his first complete line, the Grand Junction Railway from Warrington to Birmingham, which opened in 1837. The one large viaduct on this route stretched across the Vale Royal near Northwich, and its pattern of 20 regular arched spans became a prototype for hundreds of similar structures.

A moving train imparts a far more severe single live load to a bridge than any combination of road vehicles or pedestrians. But the stress on the structure of any substantial masonry-arch bridge imposed by its own dead load is comparable with, or exceeds, the stress generated by any practicable live load. As a result, the first railway bridges were all masonry arches. Stone beams would of course fail very rapidly under the impact of a train, due to their low tensile strength, but it soon became clear that cast-iron beams could be used for the many short spans that were needed. These were deemed safe up to about 12.2m (40ft); for longer distances, trusses in both timber and iron were employed, as well as cast-iron bowstring arches, with horizontal wrought-iron ties between the springings to resist the outward thrust.

The least suitable form for railway bridges was the newest: the rigid suspension bridge. Samuel Brown, whose Union Bridge over the Tweed has now carried light road traffic for 170 years, contributed a small suspension bridge for the Stockton to Darlington Railway over the River Tees, but within months it had been shaken to pieces – the weight of the locomotives far exceeding the ability of the tension structure to withstand it.

Robert Stephenson was responsible for many railway bridges, including the majestic, 28-arch, masonry Royal Border Bridge over the Tweed at Berwick, opened in 1850. In constructing this, he drove the piles into the river-bed with the new steam hammer invented by the Scottish engineer James Nasmyth, which he had first used earlier in the same year on the innovatory High Level Bridge in Newcastle-upon-Tyne. By the time the High Level Bridge was open, however, Robert Stephenson was deep into the practicalities of a yet more original work for a far more difficult site – and one already graced by another epoch-making bridge. Thomas Telford had thrown his suspended span across the Menai Strait to take the mail-coaches a quarter of a century before; now Stephenson was preparing to bridge the Strait with a far more weighty structure for the Chester & Holyhead Railway. As with Telford, it was to be his greatest achievement, and again, as with Telford, it was to be accompanied, almost simultaneously, by a smaller-scale but similar span across the Conway River near Conway Castle.

Below **The Royal Border Bridge over the Tweed at Berwick has 28 semicircular arches and, typically for railway viaducts of the time, was constructed of masonry and brick, with rubble infill. It was so named because it crosses the River Tweed close to the border between Scotland and England, and because it was opened by Queen Victoria in 1850. On her ceremonial journey to Holyrood Palace, the Queen's train was drawn by a locomotive bedecked in a livery of royal Stuart tartan. Once the Royal Border Bridge had been completed, the last gap was closed in the continuous London-Edinburgh railway line.**

programme: George Stephenson's only son, Robert (1803-1859), Isambard Kingdom Brunel (1806-1859) and Joseph Locke (1805-1860). Locke is the least-known and, from the bridge-building point of view, the least interesting of the three, but he was responsible for an immense body of economically-planned and efficiently-administered work. Beginning as an assistant to George Stephenson, he went

THE BRITANNIA RAILWAY BRIDGE, MENAI STRAIT
WALES

When Robert Stephenson confronted the challenge of carrying his Chester & Holyhead Railway across the Menai Strait, his solution – huge unprecedented wrought-iron tubes – grew out of unavoidable circumstances. After considering possible locations, he chose a site about 1.6km (1 mile) south of Thomas Telford's road bridge, where a tiny midstream island, Britannia Rock, was capable of supporting a central pier. But he was bound by the requirements of an intractable Admiralty, which insisted that there should be no centering to obstruct the waterway and no diminution of clearance at the sides of the channel. Arches, therefore, were not an option.

Even with the central tower, the two main spans of the new bridge would have to be 140m (460ft). For these, suspended decks were far too vulnerable, and solid beams would have to be of such tremendous depth that they would be impossibly heavy. Wrought-iron tubes were the only practical solution; and according to the theories of the leading authority, Professor Eaton Hodgkinson, these were going to need supporting chains. But Stephenson also consulted a ship-building firm run by a Scottish engineer, Sir William Fairbairn, which had recently built several wrought-iron railway bridges with deep

Above **Robert Stephenson**; Below **Stephenson's tubular Britannia Bridge.**

girder sides. Fairbairn carried out an exhaustive series of load tests on circular, oval and rectangular tubes, conclusively demonstrating the superior strength of the rectangular section. A one-sixth scale model supported 86 tons before it failed, disproving the need for any supporting chains, although the towers on the bridge were still given slots to hold them and built high enough to carry them.

The structural form was thus decided; now the bridge had to be built, beginning with the masonry towers and abutments, which were designed with a uniform and somewhat monumental Egyptian severity. The central Britannia Tower soars 70m (230ft) above its foundations; the Anglesey and Carnarvon Towers, on either side of the main spans, are only 3m (10ft) shorter, and their abutments, each 50.6m (176ft) long, terminate 70m (230ft) away from them. As the railway line was to be double, it needed two continuous parallel lengths of tube, each 461m (1,513ft) long; and to give the locomotives adequate clearance, the maximum external depth and width of the tubes were 9.1m (30ft) and 4.47m (14ft 8in).

All the tubes were constructed from flat plates and angle-pieces, the sections from the abutments to the side towers being riveted together *in situ* on huge

timber scaffolds. But for the 140m (460ft) centre spans there was no alternative other than to pre-fabricate each of the 1,800-ton sections and float them out successively on pontoons. Each section was 144m (472ft) long, to allow for adequate bearings inside the towers; and to accommodate this extra length during erection, the towers were built with niches in their flanks near the bases, and huge grooves running right up their inner sides to the level of the span. In June, 1849, the first tube section was guided at high tide between its niches and allowed to settle into place at the bottom of its grooves. Then steam-powered hydraulic presses near the tops of the towers took over, slowly hauling the tube up the grooves in 1.8m (6ft) lifts. During each lift, timber was continuously packed under the ends of the tube as a safety precaution – wisely, as one hydraulic press blew out – and as the tube rose, courses of brickwork were laid beneath to complete the tower-faces.

The Britannia Bridge, the last of its two million rivets driven home by Stephenson himself, was opened in March, 1850. The weightiest possible trainloads produced only tiny deflections; the design was a triumphant, if expensive, success. The Bridge continued to function until 1970, when it was severely damaged by fire. The intense heat made the tubes sag and fracture so badly that they had to be removed and replaced by a new double-deck structure on steel arches, with a road above and the rail below.

Right **A portion of one of the tubes from the Britannia Bridge. The side plates of the tubes were alternately 1.98m (6ft 6in) and 2.64m (8ft 8in) wide, and between 13mm (½in) and 16mm (⅝in) thick. The top plates all measured 1.83m (6ft) long, 0.53m (1ft 9in) wide, and were a little thicker than the side plates. The bottom plates were much larger – 3.6m (12ft) long, 0.71m (2ft 4in) wide, and from 11mm (⁷⁄₁₆in) to 14mm (⁹⁄₁₆in) thick. The tubes were strengthened by the inclusion of longitudinal cells with L-beam and T-beam dividers along both the top and bottom – 8 cells above and 6 beneath.**

Left **A contemporary print records Robert Stephenson on 5 March, 1850, inside one of the tubes, ceremonially driving home by candlelight the last of the rivets (which was subsequently painted white). On that opening day, he rode through the finished bridge on a train drawn by three locomotives and carrying a thousand people.**

BRUNEL'S TIMBER VIADUCTS

ENGLAND

Isambard Kingdom Brunel was the most gifted of all the men who led the expansion of British railways in the mid-19th century. His versatile genius extended even to the design of ships and he has also been described as England's greatest ever timber engineer.

His best-known railway was the Great Western, but it had few large bridges. It was the later South Devon, West Cornwall, and Cornwall lines which between 1849 and 1864 necessitated the construction of as many as 64 viaducts across often deep valleys. In 1859 alone, no fewer than 34 viaducts, some over 300m (1,000ft) long, were opened between Plymouth and Truro – and all of timber. In an age when engineers were exploring the possibilities of wrought iron and would soon engage with the even more dramatic structural opportunities offered by steel, Brunel's clarity of mind put fitness for purpose, economy and availability of materials first. Besides, he mistrusted cast iron for bridge spans. Patent timber trusses were carrying the new railroads of the USA across comparable obstacles, but in contrast with their repetitive designs, Brunel applied an apparently limitless inventiveness within several broad types, ranging from simple beams for short spans, through various truss configurations for longer ones, to a complex bowstring arch spanning 35m (115ft), or a near segmental shape in laminated timber slanting across the river at Bath. Supports were equally varied: timber legs could be simply vertical, outward-raking or inward-battered; and he also used masonry piers, or sometimes a combination, as at St. Pinnock in Cornwall, where splaying fingers of timber sprang from stone towers to carry the railway 46m (151ft) aloft.

They are all gone, the last in 1934, defeated by loading demands and maintenance difficulties; only old photographs, drawings and engravings now bear witness to the delicacy, grace and variety of those airborne wooden lattices.

Above **Brunel supervising another great enterprise: an attempt to launch his steamship "Great Eastern" in 1857, two years before his death.**

Below **Part-elevations of a few of Brunel's timber structures: below: a photograph of the Gover Viaduct, 210m (690ft) long and 29m (95ft) high; bottom centre: St Germans Viaduct, 288m (945ft) long and 32m (106ft) high; bottom right: St Pinnock Viaduct , 193m (633ft) long and 46m (151ft) high. All were opened in 1859 on the Plymouth-Truro section of the Cornwall Railway. Bottom left: the main 33.5m (110ft) span over the River Tawe on the Landore Viaduct on the South Wales Railway.**

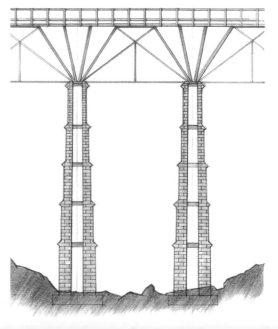

THE CLIFTON SUSPENSION BRIDGE, BRISTOL
ENGLAND

The bridge over the Avon Gorge at Bristol was not only Brunel's first major work, it was also a posthumous masterpiece. Already a veteran from having worked as resident engineer on his father Sir Marc's Thames Tunnel, the 23-year-old Brunel proposed four different suspension designs for Clifton in 1829. All were much longer than Thomas Telford's Menai span, and two exceeded even the Grand Pont Suspendu, completed five years later. However, Telford expressed doubts about their stability, and the Bridge Company called for further proposals. Brunel

Left The Clifton Suspension Bridge under construction in the early 1860s. The towers had been completed, apart from their copings, some 20 years earlier; here the first chain is about to be assembled on falsework carried by wire ropes slung between the towers.

Below **Clifton Suspension Bridge as it is today, spanning 214m (702ft) across the Avon Gorge. The bridge has required some maintenance since its completion but the wrought-iron structure has suffered little corrosion and remains largely as constructed.**

then submitted a design with the span reduced to 183m (600ft), and it won – although his final detailed design incorporated an increase in span.

Work began in 1831, but lack of money caused repeated delays. In 1835, when a recent visitor to Switzerland suggested that the timber deck be suspended from wire cables, as on the new bridge at Fribourg, Brunel defended his decision to follow Telford at the Menai Bridge and use bar chains. As years passed, and as work on the piers stuttered onwards, he rethought his designs for the deck, the chain bars and the links; and in 1840 he decided on four chains in two pairs, rather than one either side. In the same year, fabrication of the ironwork began, but in December, 1842, Brunel was instructed to cease work; and in 1849 the chain iron was sold.

Shortly after Brunel's death in 1859, his colleagues at the Institution of Civil Engineers formed a company to complete the Clifton Suspension Bridge. Fortuitously, his own Hungerford Suspension Bridge, which he had built across the Thames between 1841 and 1845, was then being demolished (although its piers remain today supporting the present railway bridge), and its four chains were purchased for re-use at Clifton. As these were lighter than had originally been planned, a third chain was added on each side, and the deck was modified: it was made of wrought iron rather than timber; was slightly narrower than on Brunel's design; and at 74.7m (245ft) it was somewhat more elevated. But when the bridge was finished at the end of 1864, it was still largely as he had finally intended.

THE ROYAL ALBERT BRIDGE, SALTASH
ENGLAND

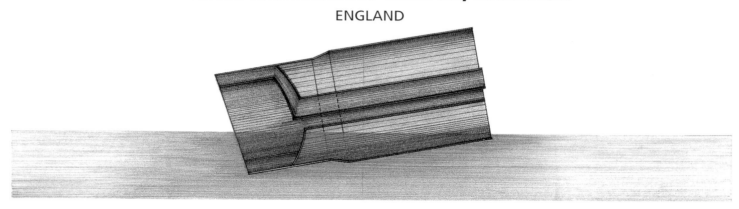

The greatest obstacle to Brunel's Cornwall Railway was the Tamar River. His first intention was to cross it with another and much bigger multi-span timber bridge, but the Admiralty's requirements for headroom and a minimum span drove him to a high-level, two-span, wrought-iron design. The biggest problem here, though, was not the spans themselves, but their middle support – the Tamar afforded Brunel no convenient outcrop on which to found the central pier; the sloping riverbed at Saltash was over 24.5m (80ft) below high water.

The answer was pneumatic sinking. Starting in the spring of 1853, he built on the river bank a huge, double, wrought-iron cylinder, 11.3m (37ft) wide and 27.4m (90ft) tall, in which a 3m (10ft) inner access tube opened out like a bell around a working chamber at the bottom. Off-centre, within the access tube, there was another, 1.8m (6ft) wide, through which compressed air was forced down into a 1.2m (4ft) wide ring around the chamber, expelling the water and keeping it out under pressure, so that the masons could work within.

The cylinder was floated out and sunk into place in June 1854. Little was known of the effects of working in compressed air, and the first workmen sent down the narrower tube under air pressure, to clear mud and rock from the annular ring and stabilize the structure, suffered the "bends" when they returned to the surface. This done, however, they were relatively free from the peril, and construction of the central pier within the chamber continued until it reached its full 30m (100ft) above high water in the autumn of 1856. After that, the two halves of the caisson were towed away, and some of the iron was saved to be reused on the bridge deck.

Founding and constructing the central pier. First the cylinder, or caisson, was floated out (above), **positioned** (below) **and then embedded. Men excavated within the annular ring until bedrock was reached, and built a 2.1m (7ft) high granite ashlar ring inside, with pig-iron stacked above to keep the caisson stable** (below right). **Then the central part was excavated, and the masonry built up to its full height** (facing page).

On an earlier bridge at Windsor, Brunel had designed a wrought-iron tied arch whose upper members had a most original V-profile; and at Chepstow, following this, he had linked two 91.4m (300ft) span wrought-iron decks, each carrying a rail track, to two 2.7m (9ft) hollow wrought-iron tubes, barely arched, one above each deck. In this way, two linked trusses had been formed, but with a slight suspicion of a suspension structure quite different in design from the familiar catenaries suspended from towers. At Saltash, on a yet larger scale, he developed this into an original suspension/arch/truss combination. For financial reasons, the bridge was to carry a single line only, with two tubes arranged linearly, curving between the three piers and supporting the deck beneath on hangers. The tubes were vast ovals, and their ends were tied by suspension chains, with a dip equal to the rise of the arches. The vertical suspension rods were braced with diagonals, and each whole assembly of arch plus chains plus hangers plus deck was prefabricated onshore, floated out on pontoons and raised into position by hydraulic presses. The first truss was floated out on 1 September, 1857, and the bridge was opened by Prince Albert on 3 May, 1859.

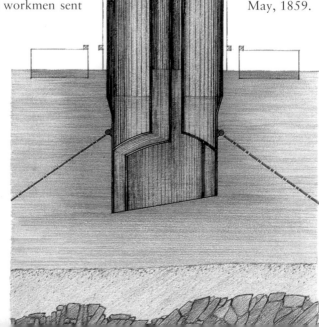

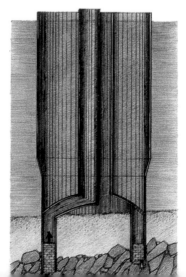

Right **The Royal Albert Bridge still carries the single line of railway across the Tamar Estuary and into Cornwall (although its glory was somewhat diminished by the completion alongside it in 1962 of the Tamar Road Bridge. At the time, this was England's longest suspension bridge, with a main span of 335m (1,100ft) and 114m (374ft) side spans, slung from 67m (220ft) reinforced concrete towers).**

Right **Raising the two trusses – which had to be done one at a time in order to keep the channel clear – took 13 months. Each span was lifted 0.9m (3ft) at a time, first one side then the other, the masonry of the shore piers being built up after each elevation while the span was temporarily supported.**

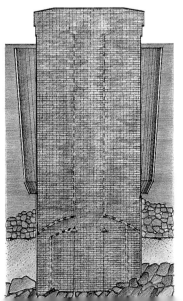

THE TAY BRIDGE DISASTER
SCOTLAND

Though railway-building in Britain peaked during the first half of the 19th century, the remainder of the century saw much continuing activity. But there was a change, specifically in bridge design. Where once pioneering and innovation had been backed up by flair and meticulous attention to detail – supremely on the Britannia and Royal Albert Bridges – there were now times when consolidation and competition eroded the highest standards of care.

Two rival railways linked Scotland and England, the North British to the east and the Caledonian to the west. South of the border, the North British had the advantage, but between Dundee and Edinburgh its line was interrupted by two huge estuaries, the Firths of Forth and Tay, forcing its passengers to break their journey for slow ferry crossings, which caused sea-sickness and which were often cancelled due to bad weather. The North British engineer, Thomas Bouch, had long argued the case for Tay and Forth Bridges, and in the 1860s, when a complex war of tactics and takeovers gave the upper hand to the Caledonian, it became increasingly clear that these two unprecedentedly-scaled bridges would have to be built if the North British was to survive.

Each bridge was to be over a mile from shore to shore, but at least the Firth of Tay was relatively shallow, and Bouch had already completed many wrought-iron railway viaducts, including one of the largest in the country, above the Belah valley in the Pennines. His design for the Tay was merely the repetition of many others, on an inflated scale.

Work began in 1871 and continued for six years, often with considerable difficulties, both in founding the piers and in the hydraulic raising of the trusses. Indeed, 20 lives were lost in several accidents, but on 26 September, 1877, the first test train, carrying the North British directors, and with Thomas Bouch on the footplate, crossed what "the Great McGonagall" was to immortalize in verse as the "Beautiful Railway Bridge of the Silv'ry Tay!" Passenger services were inaugurated on 31 May, 1878. A year later Queen Victoria herself crossed the world's longest bridge, and the "engineering triumph" earned a knighthood for its designer, whose scheme for a monster (and monstrous) suspension bridge over the Firth of Forth was already on site.

At 5pm on Sunday, 28 December, 1879, a furious gale hit the Firth of Tay, blowing full against the bridge. Two hours later, with the storm at its height, a mail train with six coaches trundled slowly on to the southern end. It never reached the north bank. For most of the line's length, trains ran above the

deep trusses, but for the central 13 spans, dubbed "The High Girders", the position of the trusses was switched so that trains ran inside them – virtually in a lattice tunnel – thereby leaving some 1,000m (3,300ft) of greater clearance for large ships at the centre of the channel. The High Girders had no braced connection with the remainder of the bridge on either side, and as the train passed through them they were blown down, complete with cast-iron columns, the train itself and 75 passengers, in the most terrible bridge and railway accident the country had ever known.

The Court of Inquiry's conclusions were merciless. Sir Thomas Bouch had been fatally complacent –

Below **The collapse of the Tay Bridge was the biggest disaster to befall Victorian engineering and was vividly documented, as in this illustration from the** *Illustrated London News*.

Above **The first Tay Bridge (bottom). Its replacement (top) was a far more substantial** structure, although its central 13 ''High Girders'' occupied the same position.

about the design, about the quality of workmanship, about maintenance and, crucially, about wind pressure. He had barely considered it, relying on century-old tables by John Smeaton, which gave as little as 12 pounds per square foot for a "tempest". From this disaster British engineers came to appreciate the significance of dynamic forces generated by violent gusts as compared with steady wind pressure. Post-Tay bridges were soon being designed to withstand wind pressures as high as 56 pounds per square foot.

THE TAY BRIDGE FACTS

First bridge	
Constructed	**1871**
Type	**single track**
Piers	**brick**
Second bridge	
Constructed	**1882-7**
Type	**twin track**
Piers	**brick/concrete/ wrought iron**
Both bridges	
total length	**3,264m/10,711ft**
number of spans	**85**
length of navigation spans	**11 of 7.47m (245ft), 2 of 69.2m (227ft)**

THE FORTH RAIL BRIDGE
SCOTLAND

"The supremest specimen of all ugliness"
William Morris

At its optimum crossing-point, at Queensferry, a few miles west of Edinburgh, the Firth of Forth is a little narrower than the Tay at Dundee, but the depth of the water – up to 65m (220ft) – made it impossible to build numerous piers for many truss spans. Only the small mid-stream Isle of Inchgarvie allowed the construction of an intermediate pier. Sir Thomas Bouch's design for the site was, understandably, dropped after the disastrous collapse of his Tay Bridge. Nevertheless, even in its short life, the Tay Bridge had been a financial success. As a result, a new bridge was built to replace it, and a new design was sought for the Forth.

By the time the new Tay Bridge was opened, in 1887, construction of the Forth Bridge was well

Below and far right
Oblique views of the Forth Bridge emphasize the sheer massiveness of the masonry piers and the main tubular steelwork, and the complexity of the girder framing that connects the principal members and encloses the deck.

under way. The designers, Sir John Fowler and Benjamin Baker, incorporated two major innovations, both pioneered elsewhere: the use of steel and the cantilever principle. James Eads had completed the first great steel bridge in St Louis in 1874 (see pp. 82-3); now, after a surprising delay, it was legal to build bridges with steel in Britain, and Fowler and Baker took full advantage of its combination of tensile and compressive strengths. At the time only one other cantilever bridge had been built for a railway, and that was in Germany, but after considering three suspension configurations, the two engineers were convinced that the "continuous girder" was the most appropriate. Haunted by the spectre of the Tay, they knew they had to build not only the largest but also the strongest, stiffest, and hence the safest, bridge in the world.

The three piers, at North and South Queensferry and on Inchgarvie, each rest on four 21.3m (70ft) circular caissons in granite-faced rubble with granite caps. The steel towers were erected on the piers, straddling from a width of 36.6m (120ft) at the base to 10m (33ft) at the top. From these, the six cantilever arms were then built out on both sides simultaneously to maintain equilibrium. The main compression members were pairs of 3.6m (12ft) diameter circular tubes springing out and up from each side of the bases of the towers in a gently flattening curve, braced by ties and struts from the straight top members projecting downwards from the tower-heads.

The bridge proper is approached by viaducts, with three masonry arches and five girder spans to the north and four masonry arches and ten girder spans to the south, both of which in any other circumstances would be major bridges in their own right. The ends of the outer cantilever arms from the North and South Queensferry piers were tied to the masonry piers at the inner ends of the viaducts, whilst from their other sides, and also outwards from both sides of the Inchgarvie cantilever, sections of the two 107m (350ft) suspended-girder spans were built bay by bay until they joined above the centre of the channels to complete the clear spans.

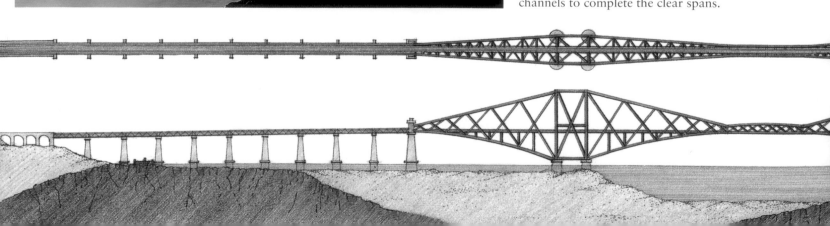

The Forth Bridge shattered records. The volumes of masonry for its piers, the height, length and depth of its cantilevers, the scale of its free spans, the volume of steel in the whole structure were all world-beaters and even today it remains one of the world's biggest and most famous bridges.

The enduring success of the Forth Rail Bridge is probably the result of its designers developing old techniques rather than risking anything new at so massive a scale. Both steel and the cantilever principle had recently been successfully employed, and giant metal tubes in compression had been applied to

Below **Only in plan and elevation do the clarity of structural form and the balance of scale between the main towers and the suspended spans become apparent. The grace of the Forth Bridge is clearly visible.**

Brunel's Chepstow and Saltash Bridges decades earlier. If Brunel and Stephenson were the more profound innovators, Fowler and Baker still produced the mightiest piece of Victorian engineering.

THE FORTH RAIL BRIDGE	FACTS
period of construction	**1882-89**
total length	**2.46km/ 1½ miles**
weight of steelwork	**58,000 tons**
clear spans	**2x521m/1,710ft**
height of towers	**100.6m/330ft**

MOVING BRIDGES: TOWER BRIDGE, LONDON
ENGLAND

Bridges do not have to be fixed. The pontoon bridges used by the armies of Darius and Xerxes were the archetypes for floating, temporary structures from ancient times to the present day; and expanding mechanical types of military bridge were being described and illustrated as early as the 16th century. The familiar drawbridge has an ancient pedigree, if Herodotus' description of Queen Nitocris' Babylonian structure really implies a lifting section; and the Hollywood image of a single wooden span rattling upwards on chains to deny marauders access over a castle moat was equally a part of reality. In the Middle Ages, drawbridges fitted with counterweights, which enabled the bascule (literally "seesaw" or "rocker") to be raised and lowered, were not only a regular feature of castles; a defensive rising

Above **One of the best-known images of London – Tower Bridge, with its central steel bascule spans raised to permit the passage of a ship upriver. In 1976 the spans were electrified, but some of the original hydraulic machinery has been preserved as part of the bridge/museum complex.**

section was frequently built into multi-span river bridges, including Old London Bridge.

A further type of moving bridge is the swing bridge, pivoting from one end or from a central pier, and many others were developed from the 19th century onwards. Some bascules were designed to move back on rollers rather than hinge upwards; lift bridges have counterweighted decks that rise vertically between towers; and jack-knife bridges have a deck that folds transversely. In a kind of opposite version of the lift bridge, decks have even occasionally been made to sink below water level to allow river traffic to pass over them. One rare, spectacular, and now quite inadequate type of moving bridge, is the transporter bridge, on which the continuous deck is replaced by a platform that moves from one side to the other, suspended from an overhead frame – as much an airborne ferry as a bridge.

Incomparably the most famous moveable bridge, however, is a drawbridge – or rather a double-leaf bascule with suspended side-spans, incorporating two high-level footbridges. Tower Bridge has become a symbol of the City of London, so familiar that it takes an effort to see it afresh as the curious hybrid that it is. Its *raison d'être*, as ever, was the pressure of traffic between the north and south of the Thames, and because at the time when it was built, 1886-94, it was the closest crossing to the mouth of the river, it was essential that it should still allow the passage of large ships. As so often happens, the form was determined by circumstances, Parliament stipulating that an opening span be provided.

The design was a collaboration between the engineer John Wolfe-Barry and the architect Sir Horace Jones, who was constrained by the other circumstance determining the form of the bridge – proximity to the Tower of London – which obliged him to provide it with a suitably "Gothic" appearance. The resulting masonry-encrusted, steel-framed towers concealed in their bases the massive hydraulic machinery to raise the twin bascules, and their tops were linked by footbridges so that pedestrians were still able to use the bridge while the roadway was being raised and lowered – a process that took about an hour.

Though loved now by a world-wide public, Tower Bridge earned the derision of aesthetes and engineers alike. Their scorn was directed not only at the architecture but also at the strange, lop-sided and ponderous suspension chains, or rather drooping curved trusses, which support the side spans. In the 1940s, Tower Bridge weathered a proposal to get rid of the Victorian Gothic cladding and clothe the structure in glass – thus replacing one generation's kitsch with another's – but in the 1980s the whole bridge was turned into a museum of itself, which in a sense it always was.

Right **The Newport Transporter Bridge over the River Usk, Gwent, South Wales. Two lattice steel towers, anchored on each side of the river, support a 197m (646ft)** span, part suspended, part cable-stayed, along which a travelling frame moves and from which is suspended the gondola carrying vehicles and pedestrians.

TOWER BRIDGE	FACTS
period of construction	1886-94
opening span	79m/260ft
side-spans	82.3m/270ft
height of footbridges	43m/141ft

7

NEW WORLD, OLD AND NEW IDEAS

The American War of Independence could be described as the trigger that fired the starting-pistol for bridge-building in the new country. Before then, the colonies on the eastern seaboard were relatively self-contained, communicating more with Europe by sea than overland.

The earliest settlers would have brought with them knowledge of the bridge-building styles and methods that prevailed in their home-countries; so it is not surprising that the only American bridges now surviving from before the Revolution are a few masonry structures in New England, such as the three semi-circular arches of the Penne Pack Bridge near Philadelphia (1697) and the two segmental spans over the Ipswich in Massachusetts (1764). But even in those days, such modest stone structures were rarities: the pioneers usually found their building materials in the abundant pine-forests.

With the establishment of the American nation, the need for good highways to carry trade between its States became quickly evident, and so too did the need for the bridges that would carry those highways across the major rivers that flowed eastward to the seaboard. The first and last sentences in a paragraph from the first *American Treatise on Bridge Architecture*, published in 1811 by Thomas Pope, make significant reading: "It is a notorious fact that there is no country of the world which is more in need of good and permanent Bridges than the United States of America . . . [although] . . . Necessity has already produced some handsome and extensive specimens of bridge-building in the United States."

In a rapidly growing country, as yet untouched by the Industrial Revolution, the need for speed of construction and the abundance of the material combined to make an irrefutable case for wooden bridges. Perhaps the first really ambitious example was designed and built by Colonel Enoch Hale in 1785, to carry a turnpike forming part of the main trading route from Boston to Montreal across the Connecticut River at Bellows Falls in Vermont. The sketchy surviving documentation indicates a continuous deck between 90m and 120m long (300-400ft), supported some 15m (50ft) above the river by a central timber pier on an islet, and braced with four sets of inclined struts. In its day, it was a great achievement, and it survived, although possibly in a rebuilt form, until 1840.

Colonel Hale's bridge was not, however, covered. This most characteristic of old American bridge profiles seems to date from 1805, when one of the most prolific and skilful of early American bridge-builders, Timothy Palmer, erected a triple-span arch-truss over the Schuylkill River in Philadelphia. Built on the site that had previously been chosen for Thomas Paine's long-span iron bridge and a subsequently abandoned proposal for a triple masonry arch, Palmer's "Permanent Bridge", consisting of

Below **The Spreuerbrücke, Lucerne, Switzerland. Such European timber-covered bridges, from the 16th century and earlier, would have been familiar to many early American settlers.**

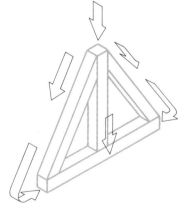

Above **The king-post truss, a basic feature of medieval timber building, reappeared in many 19th century American timber and iron bridges. It could be multiplied, extended, inverted or overlapped.**

45.7m (150ft) side-spans and a 59.5m (195ft) central clearance, would have remained an open frame had not the president of the Schuylkill Bridge Company insisted that it be completely enclosed in boarding for protection. It was a commonsense approach to the durability problems inherent in timber, and was soon followed all over the country. The creaking, dark, mossy tunnels became one of the earliest characteristics of rural pre-industrial America.

Unlike Enoch Hale, Palmer built many bridges. Indeed, he was one of the first professional bridge-builders in the USA. Other prominent names were Lewis Wernwag, who pioneered timber cantilevering in the West, and whose *chef d'œuvre* was the single-span 103.6m (340ft) "Colossus" Bridge built over the same Schuylkill River at nearby Fairmount in 1811; and Theodore Burr, who exceeded this record-breaker in 1815 with a 109.9m (360ft) timber arch over the Susquehanna at McCalls Ferry, south-west of Lancaster, Pennsylvania, which was destroyed two years later by ice.

The early American bridge-builders combined the arch and the truss in a variety of ways, without any clear understanding of truss action, and it is difficult or impossible now to analyze which of their structural features were carrying the loads and which were redundant. It is also impossible to know if any of them referred to Palladio's truss designs. But there were some corresponding features. The simplest example was an extension of the king-post, to which Burr added an arch for additional strength. Wernwag's "Colossus", on the other hand, concentrated on the arch. Within the overall covering, painted white with small windows along its length, the main

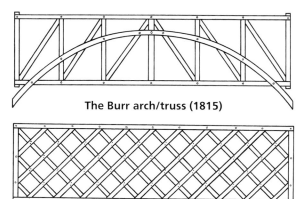

The Burr arch/truss (1815)

The Town lattice truss (1820)

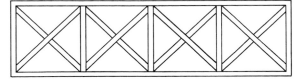

The Long truss (1830)

Above **The few American covered bridges that have survived, such as this one over the Connecticut River, Woodstock, Vermont, of the Town lattice truss type, are now preserved as treasured local and national monuments.**

curved arch beams above and below were separated by a truss composed of heavy wooden verticals and light iron diagonals – a striking similarity to the fourth of Palladio's examples (see p.34). The "Colossus", which was destroyed by fire in 1838, was something of a "one-off", but Burr's arch/truss combination was repeated all over the north-eastern States. Its drawbacks were that, on a bridge of any size, the structural members had to be large and were therefore difficult to handle with primitive resources, and their jointing required a great deal of carpentry skill.

In 1820, a Connecticut architect, Ithiel Town, invented and patented a simpler lattice truss frame which was easy to assemble and which lent itself equally to short or long spans, provided that the members of the long spans were doubled or tripled to increase their strength. Hundreds of Town truss bridges were built in New England by the Powers family, but ten years later another simple truss design was patented, based on a series of king-posts, whose diagonal bracings overlapped to become a criss-cross lattice. The brainchild of Colonel Stephen Long, it was marketed energetically in competition with Town's. By then, however, a further improvement was on the way – and so was the railroad.

AMERICAN RAILROADS AND TRUSSES

In 1840 a Massachusetts architect named William Howe patented a design which was superficially very similar to Colonel Long's wooden truss, but its difference was crucial – the vertical king-post tension members were made of cast iron. This replaced the weakest feature of Long's truss with a counterpart in a material more able (or rather, as time would show, less unable) to withstand the increased tensile forces that the live load of a train would impose – and it did so just as the railroad network was spreading.

Four years later another patent truss, the Pratt system, reversed Howe's, this time imposing the tensile forces on cast-iron diagonals sloping outward from the centre of the truss. More types followed, their evolution driven by the demands of the railroad companies for inexpensive, off-the-peg ways to carry their thousands of miles of tracks over gulfs and gullies, rivers and streams, with the maximum speed and a minimum of specialized construction skills.

In 1847 Squire Whipple published *A Work on Bridge-Building*, which for the first time brought scientific calculation into the haphazard empirical design world of the truss. Side-by-side with this increased understanding of how trusses worked came a different kind of comprehension. In the United States, as in England, a growing number of rail bridge failures demonstrated the tensile inadequacy of cast iron and the material was gradually phased out in favour of wrought iron. In the 1850s Whipple patented a bowstring type of truss in which both were used: the upper chord was entrusted to the high compressive strength of cast iron, whilst the lower chord plus the vertical and diagonal ties were of wrought iron.

Below left (top to bottom) **Three of the many truss designs invented in 19th-century America: the Howe truss (1840); the Pratt truss (1844); and one of the truss designs of Squire Whipple. (Another form that Whipple designed was based on the bowstring girder, see p.62.)**
Below **The Fink truss used below the deck on a bridge for the Northwestern Virginia Railroad over the west fork of the Monongahela, west of Clarksburg, West Virginia, c.1875. The overlapping king-post truss elements can be clearly seen, with the uprights in compression and the diagonals in tension, forming a kind of underspanned suspension system.**

Other types were originated by two engineers on the Baltimore and Ohio Railroad, Wendell Bollman and Albert Fink. Their designs were of wrought iron to cope with the ever-increasing loading from trains and were still based fundamentally on the king-post truss (see p.76); but in different ways, and to various degrees they all combined a suspension element in their structural systems.

Howe's trusses continued to be built alongside the newer types, although they now used wrought iron, and it was one of these that failed in what at the time was the worst-ever rail disaster. In 1841 Howe sold his patent to an engineer named Amasa Stone, who then formed a bridge-building company and used the design on many railroad bridges. By the early 1860s Stone had become President of the Lake Shore and Southern Michigan Railroad, and in 1865 he replaced a bridge at Ashtabula, Ohio, with the first Howe truss to be made entirely of wrought iron. Eleven years later, in a blizzard, on the night of 29 December 1876, an 11-car train drawn by two locomotives was crossing the bridge when the driver of the first heard a noise like a cannon-shot. Turning round, he saw the second locomotive start to sink behind him. He managed to drive his own "uphill" to safety on the far abutment, but the rest of the train toppled, car by car, into the abyss.

The death-toll was high and the ensuing investigations heart-searching and comprehensive. Derailment, weakened joints and failure of the wrought iron itself were suggested. Much was still unknown about the properties of materials, and more failures were to occur before a new era of safe truss design came with the widespread use of steel.

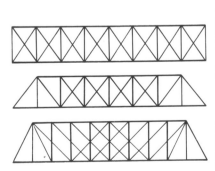

THE WHEELING SUSPENSION BRIDGE, OHIO RIVER

USA

One "new idea" from the New World had been the rigid suspension bridge, but after Finley's early structures (see p.56) the development lead passed to Europe via Britain. The lead returned to the USA, however, with the one great work of Charles Ellett (1810-1862).

After travelling widely in Europe and studying at the famous *École des Ponts et Chaussées*, 22-year-old Charles Ellet returned to his native United States in 1832. Possibly he had seen some of the early American suspension bridges, but it was what he saw in Europe that fired his life-long enthusiasm for the form. During the next few years he submitted various proposals, including two for bridges across the Potomac and the Mississippi, but his first chance to build a bridge did not come until 1838, when Lewis Wernwag's wooden "Colossus" over the Schuylkill River succumbed to fire. The 109m (358ft) Fairmount Park Suspension Bridge which Ellet completed to replace it in 1841 – America's first long-span suspension bridge and its first bridge with prefabricated wire cables (five on each side) – was only the prelude to Ellet's greatest achievement.

At 308m (1,010ft), the Wheeling Bridge over the Ohio River decisively toppled the 16-year record of the Grand Pont Suspendu in Fribourg as the world's longest span. Originally erected between 1847 and 1849, it had a 7.3m (24ft) wide deck supported on each side by six parallel wire cables, the longest ever to have been prefabricated and raised into position, rather than air-spun. Both towers are land-based with short side-anchorages, the east bank being far higher than the west. The western approach road is carried on a rising embankment, but this compensates for only part of the difference in height, so that, most unusually, the whole bridge slopes up from the west to the east, with the top of the east tower some 10m (32ft) above its counterpart. Drivers and pedestrians are thus faced with a 1:25 incline. Ellet's report on the feasibility of the bridge made clear his concern that the angle of the stays and the verticality of the towers should compensate for any imbalance arising from the slope. He allowed for expansion and contraction of the cables by letting them run freely over iron rollers on the tops of the columns.

In his Potomac proposal Ellet had theorized about the overall stability of suspension bridges and argued that their inherent flexibility, together with their increased span (and hence greater total deck-weight) would contribute to a self-maintaining equilibrium. Because these bridges presented such a small profile to lateral wind-forces, he held that they could survive high winds provided they had strongly framed deck floors; the Wheeling Bridge proved him wrong, as a violent gale blew it down in 1854.

The subsequent fame of Ellet's great rival, John Roebling, particularly his success with suspension bridge stability, has led some to denigrate Ellet's achievement, even crediting Roebling with rebuilding the Wheeling Bridge. In fact, Ellet supervised the reconstruction (although the Bridge was indeed "Roeblingized" in 1872 by John's son Washington) and it was probably only Ellet's early death in the Civil War that prevented more major bridge-building achievements. If Finley was the father of the suspension bridge in America, Ellet was the first American engineer to explore its full potential.

Below, top **The Wheeling Bridge in the 19th century, rebuilt by Ellet and showing the further diagonal bracing stays added by Roebling in 1872.** Bottom **The Wheeling Bridge, splendidly restored.**

CROSSING NIAGARA BY TRAIN
USA

In the 1830s Charles Ellet visited Niagara Gorge. He was immediately fired by the vision of throwing a great suspension bridge across it and was convinced of the practicality of his idea. In November, 1847, four months after gaining the Wheeling Bridge contract, and probably as a result, he was invited to build just such a bridge, to carry both road and railroad. In the summer of the following year, however, after he had erected a temporary suspension span as a service structure, a difference of opinion with the bridge company resulted in his resignation.

Without him, the company turned to the author of one of the other proposals for the bridge, a German immigrant called John Roebling (1806-69), who had already built several wire-suspended bridges and had

Above, inset and facing page, left **These wonderful photographs of John Roebling's first great bridge in action clearly show its unprecedented double-decked structure, and the way in which the two pairs of main cables and the stays share the load-carrying.**

set up his own wire-rope production plant. Roebling was adamant about the superiority of wire cables over chains, and expressed his views trenchantly: "There is not one good suspension bridge in Great Britain, nor will they ever succeed as long as they remain attached to their chains. . . ."

Roebling's Niagara span was less than that of the Grand Pont Suspendu or the Wheeling Bridge, but in every other respect it was a far more massive, ambitious and modern structure. The four cables each contained 3,640 wrought-iron strands, continuously air-spun in the manner pioneered by Henri Vicat. Instead of simply laying them parallel, Roebling compacted them into continuous 250mm (10in) cylinders and wrapped them overall with more wire for weather protection – a technique which he had patented as far back as 1841. To accommodate both the single railroad track and the roadway, the bridge had a double deck, with the roadway running 5.5m (18ft) beneath the rail deck, separated from it by vertical wooden members so that the whole formed a huge deep girder. Roebling's concern with overall stability did not end with the deep truss, however. The wire stays of the Niagara Bridge not only ran vertically from the cables to the deck but also fanned out from the tops of the towers, some of them terminating at the rail deck and the others extending to the roadway.

Below **The Niagara Bridge was the proving ground for Roebling's suspension system of air-spun wire cables with additional diagonal bracings. The system was equally successful on his subsequent achievements, culminating in the Brooklyn Bridge, shown here, on which the spinning of the steel strands was more efficiently mechanized than ever before.**

After the collapse of the Wheeling Bridge, he added more cables running diagonally down to the sides of the gorge from the roadway in order to eliminate any remaining chance of the hair-raising bucking that had been seen before the Wheeling's deck twisted and plunged into oblivion.

The Niagara Railroad Suspension Bridge opened in March, 1855, and was a huge success, carrying increasingly heavy traffic until necessity forced its replacement in 1897. In his final report, Roebling left a vivid picture of himself: "sitting upon a saddle on top of one of the towers of the Niagara bridge during the passage of a train, moving at the rate of five miles an hour, I feel less vibration than I do in my brick dwelling at Trenton, N.J., during the rapid transit of an express train over the New Jersey R.R., which passes my door within a distance of two hundred feet."

The same report contained another sentence whose lessons still appeared to be unlearned nearly a century later. "Undulations, caused by wind, will increase to a certain extent by their own effect, until by a steady blow a momentum of force may be produced, that may prove stronger than the cables."

THE NIAGARA BRIDGE FACTS

constructed	**1851-55**
suspended span	**250m/821ft**
height of deck above water	**74.4m/245ft**
depth of truss	**5.5m/18ft**
depth to span ratio	**1:46**

Above ***The Great International Railway Suspension Bridge***, a famous engraving of Niagara, published 1859.

JAMES EADS AND THE ST. LOUIS BRIDGE
USA

Charles Ellet's scheme of 1840 to build a bridge across the Mississippi at St. Louis was his most ambitious and also the easiest to dismiss. The river was 460m (1,500ft) wide, fast-flowing, had huge tidal range, carried massive quantities of treacherous scour in summer, and was frequently blocked by thick ice in winter. In 1840, a Mississippi bridge was impossible. At the same time, however, the man who was to make the dream a reality, James B. Eads (1820-1887), had just begun his long career on the river. After working briefly in salvage, he spent 20 years designing, building and operating numerous vessels, and during the Civil War he constructed a fleet of ironclad gunboats for the Union.

With peace came further railroad development, and the lack of a bridge to connect lines east and west increasingly crippled St. Louis' effectiveness as a trading centre. Fund-raising was launched, a bridge

Below A somewhat foreshortened contemporary view of the St Louis Bridge. It was at once the world's first really significant steel structure and the biggest bridge ever built at the time. On 1 July, 1874 James Eads test-loaded it with 14 locomotives, and on 4 July the city celebrated his triumphant structure.

company was established – in the face of strong opposition from those with vested interests in the ferries – and Eads was appointed its engineer-in-chief. He had never built a bridge before, and this was to be his only one, but he knew the Mississippi intimately – particularly its deep and treacherously shifting bed of sand. The spans would have to be founded to bedrock, at an unknown depth.

In August, 1867, while still battling with the opposition, Eads began work on the West Abutment. Here, the bedrock was nearest to the surface, but multiple layers of wreckage and industrial waste hindered the progress of his cofferdam. In the spring of 1868, when the slow work was halted by flooding and the doubts were reinforced, Eads saved his scheme by publishing a brilliantly clear exposition of it. Apart from the difficulty of founding piers, he was obliged to allow sufficient clearance and unrestricted

passage for river-boats, so he proposed a double-deck-road above, railroad below, three-arch structure in which the size of the arches was unprecedented when built: they spanned 153m, 158.5m and 153m (502ft, 520ft and 502ft). Also, this was to be the first major bridge built in steel, for which inexpensive methods of mass production had only recently been discovered.

Congress granted approval, and the West Abutment was completed; but this was only the beginning. The bedrock sloped down from west to east, and Eads decided to use pneumatic caissons for the two piers and the East Abutment, having observed the method on bridge foundations at Vichy, France. The East Abutment would be the deepest, and most difficult, so he began with the piers. On the East Pier a borehole test had passed through 4.3m (14ft) of water and 25m (82ft) of mud before striking bedrock. On 28 February, 1870, the East Pier caisson touched bedrock, but already the air-pressure at depth had caused many cases of caisson sickness, and shortly afterwards came its first fatalities: 12 men died from "the bends" on the East Pier alone.

The shallower West Pier reached bedrock that April, and then Eads turned to the East Abutment. By then his own physician, Dr. Jaminet, who had suffered severely from "the bends" himself while treating East Pier victims, had concluded that slow decompression was the answer (as was already known in Britain and France). Other improvements were implemented, including an elevator in the East Abutment caisson and much better rest facilities. As a result, although bedrock was struck at no less than 41.5m (136ft) below high water, only one death from "the bends" occurred on the East Abutment. After surviving a tornado in the final stages of construction, it was completed early in 1871, with a volume five times that of Brunel's single Saltash pier.

After 3½ years' work, Eads' foundations were ready to receive his great steel arches: his material specification was unprecedented in both its scale and the quality of workmanship demanded, setting a benchmark for future standards. Construction proceeded by cantilevering via falsework from the piers and abutments. The arch-halves were closed in early 1874, and the bridge opened to pedestrians in May.

THE ST. LOUIS BRIDGE	FACTS
constructed	1867-1874
type of structure	tubular steel arches
pier construction	granite-faced limestone
maximum span	158.5m/520ft
highest water clearance	16.8m/55ft
total cost	$6,536,729

Below **One of the caissons designed by Eads to found the piers and abutments of the St Louis Bridge. The iron-cased timber caisson was floated into position and sunk at the river-bed and Eads used a sand-pump he had invented to remove the huge volume of fill excavated in the compressed air-chamber. While the men worked, the granite pier being built above them gradually pressed the caisson ever deeper into the mud. The masonry was built hollow, to be filled with concrete when the caisson finally settled on bedrock.**

THE BROOKLYN BRIDGE, NEW YORK
USA

Suspension spans 1
For later spans
6-8 see p.148,
and 9-11 see p. 162.
Right **Suspension bridge
spans covered ever-
greater distances during
the 19th century:
1 James Finley's patent
design, 1810, 61m
(200ft). (see p.56)
2 Thomas Telford's
Menai Straits Bridge,
1826, 176.4m (579ft).
(see pp.58-9)**

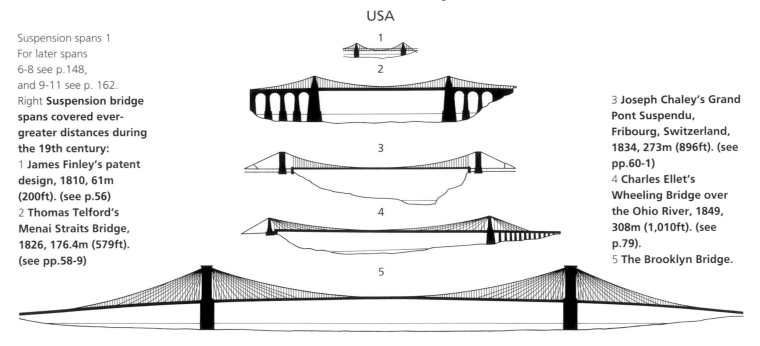

3 **Joseph Chaley's Grand
Pont Suspendu,
Fribourg, Switzerland,
1834, 273m (896ft). (see
pp.60-1)
4 Charles Ellet's
Wheeling Bridge over
the Ohio River, 1849,
308m (1,010ft). (see
p.79).
5 The Brooklyn Bridge.**

Before John Roebling began work on the Niagara Bridge, he had been engaged to survey the prospects for one across the Ohio River at Cincinnati. He reported: "The construction of suspension bridges is now so well understood that no competent builder will hesitate to resort to spans of fifteen hundred feet and more . . ." His final design proved it.

The Cincinnati Bridge was a saga protracted by opposition, financial problems and the Civil War, but it was eventually completed in December, 1866, with what was then the world's longest span of 322m (1,057ft). Almost simultaneously, vital financial weight and political will were finally united to drive Roebling's "fifteen hundred feet" span into reality, across New York's East River to link Brooklyn and Manhattan Island. His report was brilliant and comprehensive: a thorough justification of the span length (at 486m/1,595ft half as long again as the Cincinnati); all aspects of the suspension structure, including the use of steel cables and an intricate web of radiating stays; the use of deep stiffening trusses; the architectural design of the towers; and, as well as the road and the railroad, the provision of an upper, central, pedestrian promenade. Tragically, in the summer of 1869, with every detail planned, an accident while surveying the site cost Roebling his life.

His son, Col. Washington Roebling (1837-1926), chief assistant on both the Cincinnati Bridge and another across the Allegheny River at Pittsburgh, took charge. As at St. Louis, the first great task was to found and build the piers, and Roebling followed Eads in using pneumatic caissons. The bedrock proved not to be so deep – 13.6m (44ft) for the

Facing page, inset **John Roebling's Brooklyn Bridge (completed 1883). His most ambitious structure prior to the Brooklyn, the Cincinnati Bridge had a main span only 14.3m (47ft) greater than that of Ellet's Wheeling Bridge, but its massive 70m (230ft) towers and extensive side-spans made it a much bolder bridge overall. In 1869, the year work began on the Brooklyn, a slender record-breaking span of 386m (1,268ft) was completed at Niagara by Samuel Keefer. However, this did not diminish the mighty leap in scale represented by the Brooklyn: the Roeblings' great bridge still proudly dominates its surroundings more than a century after its triumphant opening on 24 May, 1883.**

Brooklyn caisson, 23.8m (78ft) for the New York – but what they were required to cut through made sinking them even more appallingly difficult. The Brooklyn side had very dense compacted clay, studded with boulders, all of which had to be hacked out from under not only the outer cutting edge of the caisson, but also the walls that divided it into large chambers – and the work had to be carefully co-ordinated so that the huge structure descended uniformly. From early summer 1870 onwards, progress was a painfully slow 150mm (6in) per week. The boulders became more numerous as the caisson went down, and at last Roebling was forced to use blasting – immensely risky in the confined space.

The Brooklyn caisson did not go deep enough for the level of air compression to become dangerous, but otherwise conditions inside were a physical and psychological nightmare. One Sunday, when no crew was working, the compressed air blew out in a huge explosion, flooding the caisson, which dropped several inches. So strong was Roebling's timber structure that it withstood the impact of nearly 18,000 tons of masonry without serious damage.

The New York pier was even more difficult. The fill was deeper and varied – foul-smelling garbage and quicksand were just two of the layers to be cut through – and the depth meant caisson sickness. Dr. Jaminet's conclusion at St. Louis about the need for slow decompression had not reached New York, and illness and fatalities. Roebling himself fell victim in the summer of 1872. Partially paralyzed for life, he was thereafter only able to direct the work from his apartment via his wife.

Right **The Brooklyn Bridge under construction in 1877. The towers had been completed in 1876, and cable-spinning was about to begin. Temporary "traveller" ropes were drawn across, and a catwalk spanned the river to Manhattan. The wrought-iron bar chains in the foreground were to connect the suspension wires to masonry anchorages.**

SAFE FOR ONLY 25 MEN AT ONE TIME. DO NOT WALK CLOSE TOGETHER NOR RUN, JUMP OR TROT. BREAK STEP! W.A.Roebling *Engr in Chg*

Right **A vivid contemporary image of excavation in progress beneath one of the caissons. Despite the massive timber platform above their heads, the crew's worst fear was of being crushed by the vast and growing mass of masonry stacked up above them on the platform.**

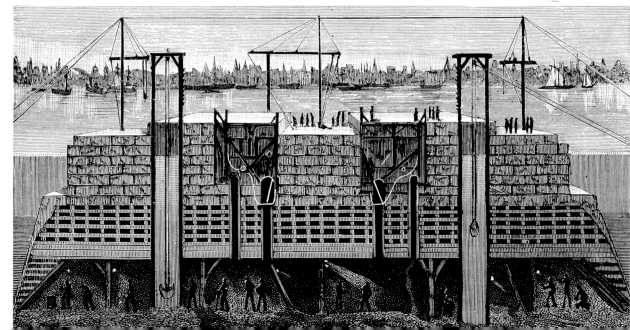

part three

In the 20th century, designers of steel arch, suspension and cantilever bridges have surpassed even the mighty achievements of their predecessors. The new technologies of reinforced and prestressed concrete have allowed an unprecedented expansion of smaller-scale bridge-building worldwide, while the advent of cable-staying after World War 2 opened up a new world of design possibilities for inventive engineers.

The George Washington Bridge, New York: probably the greatest single leap forward in 20th-century suspension bridge design.

8

THE MANY SHAPES OF 20TH-CENTURY STEEL

The three great steel bridges built between 1870 and 1890 – the triple-arch St. Louis, the suspension Brooklyn, and the double-cantilever Forth – signalled the beginning of an epoch that continues today. Nevertheless, they marked neither the first use in bridge design of steel nor the immediate end of wrought iron. Indeed, the latter was to have a glorious swan-song.

What made steel so desirable as a structural material? Cast iron – hard and thus strong in compression, but brittle and hence relatively weak in tension – includes about 3 percent carbon. Wrought iron, first made in large quantities in the 1780s by Henry Cort, has virtually the opposite properties, as Cort's puddling process drove almost all the carbon out of the pig iron, leaving a relatively soft and malleable working material. Steels – there are many – are essentially wrought iron with certain controlled proportions of carbon replaced, together with small amounts of materials like chromium, nickel and manganese, variously added to impart specific properties. In general, steel combines the advantages of cast and wrought iron without the disadvantages – but before about 1850 it could not be produced in enough quantity for large-scale industrial use.

An Austrian engineer, Ignaz von Mitis, was the first to use steel in a bridge – for the hanging eyebars of a slender suspended span over Vienna's Danube Canal in 1828, but this was rare before the mid-century flurry of new steel-smelting processes, initiated in the USA by William Kelly and in England by Henry Bessemer and Robert Mushet. In the

Above: **The climax to Gustave Eiffel's bridge-building career, the Garabit Viaduct, near St. Flour in the South of France. The two halves of the arch were cantilevered out from the piers and the side-spans. The trussed deck carried a single track railway.**

1860s, William Siemens and the Martin brothers introduced improvements to the processes in England and France; and the same decade saw the first small steel bridges completed in Europe, although at first it was difficult to control the quality. In 1865, however, Julius Baur patented chrome steel in the USA, and it was this alloy that Eads used for the St. Louis Bridge. His quality and quantity requirements were so unprecedentedly rigorous that negotiations for supply were both protracted and acrimonious, but they pulled bridge-building steel out of its infancy and changed it from wrought iron's uncertain sideline into an industry in its own right.

The first all-steel bridge came in 1879, five years after the completion of the St. Louis, when another pioneering American engineer, General William Sooy Smith, used a new type of steel, developed by A. T. Hay, to build five 94.8m (311ft) Whipple trusses to carry a railroad across the river at Glasgow, Missouri. During construction, part of the timber scaffolding collapsed, and one whole span crashed into the river. When it was recovered, it was found to exhibit not the slightest sign of brittle fracture. Even so, the Missouri Bridge did not usher in a general move away from wrought iron.

In France in particular, spectacular structures continued to be erected in the longer-established material. From 1864 onwards, Gustave Eiffel (1832-1923) engineered some half-a-dozen huge wrought-iron viaducts in the Massif Central. As the whole region is subject to high winds, Eiffel used continuous open truss spans on splayed piers, also made of open ironwork, usually with tubular corner columns in order to reduce wind resistance. In 1875,

he won a competition for a railway bridge over the Douro River near Oporto in Portugal. As before, his supports were splayed, but the central span consisted of a 160m (525ft) wrought-iron arch with a rise of 42.5m (139ft). Completed in 1877, the Pia Maria Bridge was the prelude to his bridge masterpiece, the Garabit Viaduct completed in 1884 over the Truyère River, near St. Flour in the south of France. The span, then the longest arch in the world, was 5m (16ft) longer than its Portugese counterpart, but it was much more elevated due to the steepness of the gorge, rising 122m (400ft) above the valley floor. The Garabit Viaduct was Eiffel's final bridge; a few years later he designed his most famous work – the great iron Tower in Paris, built to commemorate the centenary of the Revolution. The 1880s, of course, also saw the construction of the Forth Bridge, which at last marked the end of the quarter-century "false

dawn" of steel bridges, although a few striking structures continued to be built even in cast iron.

In 1897 John Roebling's pioneering Niagara suspension bridge was replaced by a steel arch, and in the following year a new record for steel arches was set by the nearby 256m (840ft) Niagara-Clifton bridge, which was to succumb to an ice-jam in 1938. In France, the first large steel bridge was the Viaur Viaduct, completed in 1898, with a central 220m (721ft) cantilever span; and one of the most enterprising structures from the beginning of the 20th century was built in the heart of Africa. The site was even more spectacular than Niagara – the Zambezi Gorge below Victoria Falls. Here, in 1907, an English engineer named Ralph Freeman built out the two halves of a 152m (500ft) steel arch from the sides of the Gorge, tied back by cables until they met and were connected in the middle.

DISASTER ON DISASTER: THE QUEBEC BRIDGE
CANADA

James Eads' chief engineer at St Louis was Theodore Cooper, who went on to become one of the United States of America's most eminent bridge designers. In 1899, after receiving a commission to design a railway bridge over the St Lawrence River at Quebec, he proposed a steel cantilever spanning 488m (1,600ft) with 183m (600ft) side-spans. Subsequently, however, he increased the span by 61m (200ft), bringing the piers further shoreward in order to reduce their depth and cost. This made it the world's longest span, but it was also the first link in a fatal chain of events, created not only by the cost-cutting, which reduced research and materials to a minimum, but also by the absence of the now old and infirm designer, who was far away in New York, and by inadequate supervision on the part of most of those who had been left in charge.

Work began in 1904, and by the beginning of 1907 the piers had been founded, the shoreward south anchor span had been erected on falsework, and its balancing cantilever arm had been built out over the St Lawrence. The falsework then went to the north bank, and the southern half of the central span began to finger outwards. In August, 1907, however, with the cantilever stretching over 230m (750ft), slight buckling was noticed in plates on the main beams near the base of the tower, which, like all the supporting chords, were rectangular in section, not tubular like those on the Forth Bridge. Cooper's able site engineer informed him in New York, but his order to the company to cease work pending investigation was never passed on to the bridge. At

Below left **The collapse of the centre span of the second Quebec Bridge, on 12 September 1916.**
Below right **9,000 tons of mangled steel – all that remained of the southern cantilever arm of the first Quebec Bridge, after the catastrophe of 29 August 1907.**
Bottom **A year after the second disaster a new centre span was successfully raised and the Quebec Bridge was completed in its present form. It remains the world's longest cantilever span, but is aesthetically far inferior to the Forth Bridge (in Scotland) which it superficially resembles.**

5.15pm, on 29 August, the entire structure collapsed. Eighty-five men, all due off-site in only 15 minutes, went down with it, and only 11 of them survived.

The inquiry found that there were flaws in the design dating back to the project's inception; the ninth panels in the supporting chords on both sides of the bridge had failed virtually simultaneously, as a result of the inadequate riveting of the joining plates, even though buckling had already been noticed in adjacent panels as well. Cooper's career was over, and he died a few years later.

Surviving photographs of the incomplete bridge before the disaster show an extraordinarily fragile-looking structure compared with the Forth Bridge. Its replacement used two and a half times as much steel and substituted massive, straight, although still rectangular, chords for the insubstantial curves of Cooper's design. The suspended span, however, was still much bigger in proportion to the towers than on the Forth Bridge – at 195m (640ft), it was nearly twice as long. The whole 5,200 ton structure was prefabricated, and in September, 1916, jacks began to raise it into place between the cantilever arms 46m (150ft) above the water, but when the structure had only risen 3.6m (12ft), a casting on one stirrup failed and it crashed into the water. Eleven more deaths were added to the Quebec Bridge's toll.

THE CARQUINEZ STRAIT BRIDGE, SAN FRANCISCO

USA

Gustav Lindenthal (1850-1935) was the next great American bridge designer after the Roeblings (see pp.80-1), and his Queensboro Bridge in New York, which was contemporary with the Quebec Bridge, was the first really big cantilever bridge constructed in the United States. A continuous structure of two unequal main spans of 360m (1,182ft), with a central, 192m (630ft) span and two outer anchor spans of 143m (469ft) and 140m (459ft), totalling 1,135m (3,724ft), it was unique in that it has no suspended sections: instead, the cantilevers themselves met over the water. The Queensboro Bridge was designed to carry heavier loads than any previous bridge, and it was the first one to be constructed in high-strength nickel steel.

Twenty years later, on the other side of the United States, the first Carquinez Strait Bridge at San Francisco, designed by Lindenthal's one-time assistant David Steinman (1887-1960), pushed back the frontiers still further. This is a double cantilever about the same size as the Queensboro, with two

Bottom left The central anchor span and one of the cantilever spans of the Queensboro Bridge.
Below centre The Carquinez Strait Bridge with its clearly delineated suspended sections.
Below right In 1958, this second Carquinez Strait Bridge, virtually identical to the first, was constructed about 60m (200ft) away.

main spans of 335m (1,100ft) each. The suspended sections are 132m (433ft) long and weigh 650 tons each; and there is no better evidence of advancing expertise than the fact that each was lifted into place in about half an hour using huge sand-filled boxes as counterweights, whereas the final, successful jacking of the Quebec Bridge took four days.

The Carquinez Strait Bridge also required deeper central piers than ever before – 45m (150ft) to bedrock. For safety, the caissons were made of timber, but they also had to be lined with tarred felt to protect them against the depredations of teredo, or shipworm. Then, as a long-term protection against the earthquakes for which San Francisco is notorious, the whole structure was tied into a single unit, allowing movement between sections of the bridge only to the limited extent required for thermal expansion and contraction. The first Carquinez Strait Bridge has withstood all shocks to date.

THE HOWRAH RIVER BRIDGE, CALCUTTA
INDIA

The largest cantilever bridge to be constructed in the first half of the 20th century was built in India in the 1940s by British engineers. Curiously, until the completion of the Forth Bridge, India had also contained the world's longest cantilever span, the Lansdowne Bridge over the River Indus. Now, 60 years later, the firm Rendel, Palmer & Tritton designed the immense single span of 457m (1,500ft) which today crosses the Hooghly River at Calcutta, with a suspended central section of 172m (564ft). The anchor spans are each 143m (468ft) and, unusually, not only the outer piers, which tie the ends of these, but also the piers supporting the towers are built on land. The piers were the largest ever sunk at the time, consisting of more than 40,000 tons of concrete each and requiring enormous excavations, measuring 55 x 24m (180 x 80ft) in plan and going down over 30m (100ft) through silt, sand, and the accumulated debris of many centuries.

The superstructure of the bridge was erected by

The Howrah Bridge was built to replace a floating timber pontoon bridge which had been installed in 1874 with a design life of only 25 years. The constant flow across it of pedestrians, carts, livestock, and vehicles of all shapes and sizes made a new bridge an urgent necessity.

creeper cranes. Beginning with the anchor arms on supporting falsework, the cranes moved slowly inwards from the anchor piers towards the half-built towers, and as they advanced they rose gradually higher and higher atop the leading edges of the arms. When they reached almost the full extent of the arms, high above the unfinished towers, they built up the towers to meet them at their full height of 82m (270ft), and then moved over the tops of them to the river side. From there, they continued their advance towards each other, assembling the cantilever spans and then the two halves of the suspended section out over the water (using the method that had been such a success on the Forth Bridge and a failure on the first Quebec Bridge). Finally, when a gap of only 450mm (18in) remained, the halves of the suspended section were joined by jacking them towards each other from the joints with the cantilever arms. After completion of the main structure, the deck was concreted, and the bridge opened in February 1943.

Right **The flood of pedestrian, animal and vehicular traffic continues across and beneath one of the last, and most grandiose, engineering legacies left by the British in the final years of their rule in India. The deck comprises a 21.6m (71ft) wide roadway with 4.6m (15ft) pedestrian footways cantilevered out on each side. Most of the 26,500 tons of steel used in its superstructure were manufactured in India, although many of the more intricate components were shipped out from England. Three years after it opened, one day's traffic was recorded: 121,100 pedestrians, 27,400 vehicles, and 2,997 cattle.**

HELL GATE BRIDGE, EAST RIVER, NEW YORK
USA

Gustav Lindenthal's most ambitious bridge was never built, but it would have exceeded most suspension bridges in span and dwarfed them all in breadth and weight. He first proposed it in 1899, to carry multiple rail tracks, trolley tracks, bus lanes and 16 vehicle lanes on two levels across the Hudson River, and for the rest of his long life he continued to promote it. Ultimately, however, all the traffic it was designed to take went into a complex of tunnels, and Lindenthal's greatest monument proved to be neither a suspension bridge nor a cantilever. Instead, it was a colossal arch – the Hell Gate Bridge, which links Long Island City and Ward's Island across an arm of the East River – on the other side of Manhattan Island from the site of his unrealized vision.

There are two basic configurations for long-span metal arch bridges: in one, the roadway is supported from below, which is the usual way of crossing a ravine or a gorge; in the other, the arch springs from relatively low-lying abutments and soars over the deck, which hangs from its soffit. The Hell Gate Bridge, completed in 1916, is of the second type. Carrying four tracks, it was one of the last great bridges to be built for the American railroad, and as well as being the longest arch, for many years it could carry a greater weight of continuous live load than any other bridge.

The Hell Gate's commanding site required a design that deliberately emphasized its prominence. Although almost all the loads are carried by the lower chord into the abutments, there are wide braced spandrels above, which add to the visual scale, and the curve of the upper chord reverses gently into huge masonry towers, further accentuating the overall massiveness and at the same time

Above **The Hell Gate Bridge structurally complete, but with the rail line from the approach viaduct not yet connected.**
Facing Page **Although this mighty steel arch was a triumph of early 20th century technology, two aspects of it can be seen as belonging more to the 19th century: the architectural treatment of the towers with their arches, balustrades, and elaborate cornices (below) and the very transport system that it supported** (above). **Here, the bridge and its railroad approaches hold unchallenged dominion over the flat expanse of Ward's Island, but a parallel road suspension bridge with approaches was later built alongside, signalling the ascendancy of the automobile and the eclipse of the railroad.**

creating the illusion that it is a structural necessity designed to carry some of the forces.

The most difficult aspect of the construction was founding the Ward's Island tower. In the bedrock where the supporting caissons were to be located, site investigation revealed a fault filled with clay and small rocks, which was of unknown depth and up to 18m (60ft) wide. The result was an extraordinary feat of *underground* bridge engineering. At a depth of 21m (70ft), a concrete arch was erected across the fault to carry one of the caissons, and elsewhere the adjoining cutting edges of two other caissons were cantilevered out across the crevice.

Construction of the arch itself had to be executed without falsework, so the two halves were cantilevered out from each shore. As they came close to completion, hydraulic jacks on top of the towers held the span halves 57cm (22in) above their final position. The measurement between them was found to be within 8mm (5/16in) of that calculated – fantastically accurate considering the scale of the structure – and the cantilevers were subsequently released into position to form the self-supporting arch. With the completion of the arch, the 28.3m (93ft) primary floor beam structure was suspended, and the remainder of the steel and concrete deck laid. Finally, the granite-faced towers were completed.

HELL GATE BRIDGE	FACTS
constructed	1912-16
span of arch	298m/977ft
height of towers	76m/250ft
weight of steelwork	39,200 tons
height of deck	41m/135ft
height of arch	93m/305ft

AUSTRALIAN AND AMERICAN RIVALRY: SYDNEY AND NEW YORK

The Hell Gate Bridge in New York held its record as the world's longest arch for 15 years; and when the record was finally exceeded, the circumstances involved a certain ironic paradox. Two immense steel-arch bridges of almost exactly the same span were opened within months of each other, but the one which closely followed the Hell Gate model in all but one important respect was constructed on the other side of the world, whilst its contemporary, a far lighter and more forward-looking structure, was built only a few miles from the Hell Gate on the other side of New York.

Sydney Harbour has often been described as the most beautiful harbour in the world. Proposals for a bridge across it had been considered many times before when, in 1922, detailed plans for a cantilever structure were proposed (subsequently changed to an arch when the proposer himself visited New York and came home impressed by the Hell Gate). The New South Wales Government put the project out to international tender, and although many different designs were considered, the successful English company, Dorman Long, and their designer-consultant, Ralph Freeman, settled on an arch that was close to the great American model. Despite this similarity, however, the project still represented a chance for the Australians to gain the world's longest arch-bridge and for the British to regain some of their former pre-eminence in long-span bridge design.

The result, constructed by the cantilever method with the half-arches held back by steel cables, is undoubtedly the most celebrated and the most massive

Below **Ammann's great Bayonne arch has clarity and elegance, but the open steel meshes at the abutments seem insubstantial and fail to provide an adequate visual termination to the bases of the arch.**
Facing page **Sydney Harbour Bridge has, arguably, the most dramatic and beautiful site of any major bridge in the world** (top). **Nearby, and equally prominent, is the fantastic roof profile of Sydney Opera House** (bottom), **erected in the 1960s. These two different but equally seminal structures complement both each other and the splendour of their location.**

bridge of its kind. Four rail tracks and six lanes of roadway running side by side made it the widest-ever long-span bridge at 49m (160ft). The towers, unmistakably 20th-century in their somewhat Art Deco profile, and separated into pairs at each end by the bridge's width, are a fascinating contrast to the 19th-century Germanic massiveness of the Hell Gate's. The fact that the upper chord of the arch terminates before the towers means that, visually, they punch downwards and emphasize the true springing-point of the arch's lower chord from the abutments, unlike the towers at the Hell Gate Bridge, which are physically connected to the arch and thereby have an apparent but spurious structural role.

Biggest, widest, most famous, but not the longest – Sydney Harbour Bridge had already been pipped at the post when it opened in March, 1932. The Bayonne Bridge, a mere 0.6m (2ft) greater in span, had opened in November, 1931. This bridge, crossing the Kill Van Kull between Newark and Staten Island, was the work of Othmar Ammann (1879-1966), the second of Gustav Lindenthal's assistants to become pre-eminent among 20th-century American bridge designers (the other was David Steinman). The brief here was for a road bridge, and since this required a far lighter live load than the Hell Gate or Sydney Harbour bridges, it enabled Ammann to achieve a much airier and more slender structure – an effect which he enhanced by avoiding the use of towers at each end. The upper chord of the arch continues down through the line of the deck rather than turning upwards and terminating above; the main thrusts from the lower chord on either side are carried into abutments, and over each of these there is no monumental masonry, merely a frame of light steel members ending at deck level which Ammann originally intended to clad in granite to give a solid appearance, but which in the end, he left.

SYDNEY AND BAYONNE FACTS	
constructed	
Sydney	1924-32
Bayonne	1928-31
weight of steel	
Sydney	38,390 tons
Bayonne	16,520 tons
main span	
Sydney	503m/1,650ft
Bayonne	503.6m/1,652ft
clearance	
Sydney	52.4m/172ft
Bayonne	45.7m/150ft

RENAISSANCE OF THE BOX-GIRDER
EUROPE

The reconstruction programme which followed the end of the Second World War inevitably included a vast amount of bridge work, not only replacing those that had been destroyed but also building new ones; and to meet the demands of changing uses and the constraints of limited funds and materials, new techniques were evolved and old ideas were brought back and enhanced. One remarkable resurgence was that of the box-girder, a century after Stephenson had invented it at Menai and Conway. However, the modern counterparts of those great, thick-walled, iron tubes, assembled and connected by hundreds of thousands of rivets, were far different.

As steel was scarce, and as the new generation of box-girders did not have railways or roadways running through them, the massiveness of earlier examples were neither economical nor necessary; and the erection of much lighter structures had been made possible by two "spin-offs" from wartime technology. The first was a greater understanding of the behaviour of thin-walled boxes and tubes in torsion (twisting) – what had been learned from the fuselages of fighters and bombers could be effectively transferred to tubular bridges. The second was the much improved and extended arc welding of steel sections, in which the necessary heat is generated by electricity. On the new bridges, the decks were formed out of a series of boxes welded together, with the horizontal plates (flanges) stiffened along their lengths for compressive and tensile strength, and with the side-plates (webs) either vertical, or sometimes in windy locations, angled outwards from the base to create aerodynamic stability by forming a

Below **The Cologne-Deutz bridge, with spans of 132m (433ft), 184m (604ft), and 121m (397ft), built on the piers of an earlier suspension bridge in 1946-48. A second bridge of exactly the same profile, but in prestressed concrete, was constructed alongside in 1976-80. In the background towers Cologne Cathedral.**
Facing page **Following the collapse of one of the spans (above land) of the Milford Haven Bridge, repairs were made and work continued. Here, the suspended section of the 213.4m (700ft) navigation span has been floated into position and is about to be raised.**

trapezoidal rather than a rectangular cross-section.

The post-war need for replacement bridges was greater in West Germany than anywhere else, and that country led in developing these lightweight, simple and easily analyzed and erected structures. The first long-span bridges of this type were the Cologne-Deutz and Düsseldorf-Neuss bridges across the Rhine, completed in 1948 and 1951, the latter with two 103m (338ft) side-spans flanking a central span of 206m (676ft). A particularly striking example, with a main span reaching 261m (856ft), was the Sava Road Bridge, built in 1956 in Belgrade, in what was then Yugoslavia; and in the next 15 years or so many others of similar size were constructed all over the world.

Typically, pre-assembled complete box-sections were craned into position and welded together sequentially from the heads of piers, so that they cantilevered outwards. Some were completed as true cantilevers, with a whole central suspended span raised into position; and on others the box-sections were built out from each side until they met to form the complete span. Many long-span, box-girder bridges were, and still are, assembled by these methods, but on 2 June, 1970, disaster struck on a new steel, trapezoidal, box-girder bridge at Milford Haven in Wales. A diaphragm in one box over a pier suddenly failed, and a partially completed, unsupported span collapsed, killing four men. Only three months later, a similar but completed span on the West Gate Bridge over the River Yarra at Melbourne in Australia split apart at the central joint between two box sections. Here the human tragedy was even worse – 35 men lost their lives.

These disasters, together with a third similar collapse in Europe, necessitated a thorough re-examination of the construction techniques employed on this type of bridge, which in turn led to increased understanding both of the patterns of forces on the plates forming the trapezoidal boxes and of the way in which brittle fracture of the steel itself can result from some welding techniques.

STEEL BOX-GIRDERS FACTS
advantages **strength (torsional stiffness)**
economical in material
lightweight
quick to construct
disadvantages **buckling can occur if structure not thoroughly analyzed**
can be unaesthetic

9

STEEL SUSPENSION BRIDGES

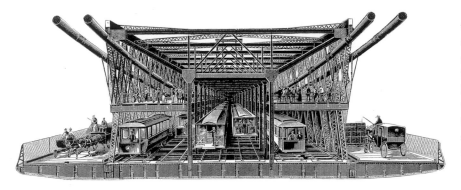

Right **A section through the deck of the Williamsburg Bridge, New York, designed by Leffert L. Buck. Four 473mm (18⅝in) steel wire cables carry a 12.2m (40ft) stiffening truss, which was one of** the deepest ever constructed in a suspension bridge. Its two decks carried rail lines, tramways and walkways, as well as vehicular carriageways cantilevered beyond the truss.

During the first half of the 20th century, an un-paralled conjunction of need, will and resources ensured that the greatest activity in long-span bridge-building took place in the USA, supremely in the development of the suspension principle for the very longest distances. In this, the eclipse of the railroad by road traffic played an important part. Despite the success of Roebling's Niagara and Brooklyn Bridges, the extra rigidity of arches, trusses and cantilevers made them more immediately suitable for railroads. The mushrooming need for road bridges in the early 20th century left behind such extreme concentrations of load, allowing enterprising designers of suspension bridges to extend the limits of the possible, both in span length and in the ratio between the span and the depth of the deck, although ultimately not without cost.

Below **The original concept for the Manhattan Bridge was for an eyebar chain bridge, but it was built as a wire cable. The span was a little shorter than the Brooklyn (behind) and the Williamsburg, but incorporated some advances – the erection of the steel towers and the air-spinning of the wire cables were more thoroughly mechanized than ever before.**

For nearly 50 years after the completion of the Brooklyn Bridge these limits were nudged gently forward. Of all American cities, New York gave the greatest opportunities to bridge-builders, as it needed many links between its separate parts across the confluence of wide waterways. The Brooklyn was just the first great long-span bridge here. Within 20 years it had a companion, the Williamsburg, less than a couple of miles north-east round a bend of the East River. Completed in 1903, it spanned 487.5m (1,600ft), only very slightly more than the Brooklyn. The Williamsburg was the first large suspension bridge with steel towers, but otherwise it did not represent a design advance.

Thereafter, the trend towards slenderness, grace and economy gathered momentum. Gustav Lindenthal, who, as city bridge commissioner, had super-

Facing page **Two views of the Philadelphia-Camden suspension Bridge across the Delaware River: top, the partly completed bridge looking from the Camden Tower across to the city of Philadelphia; below, the first crossing of the bridge in 1926. For just three years, between 1926 and 1929, the Philadelphia-Camden Bridge was the world's longest suspension span, but more remarkable was the 39m (128ft) wide deck, carrying trains as well as road traffic.**

vised completion of the Williamsburg, designed New York's next major suspension bridge. This was the Manhattan (1909), giving New York's East River three suspension crossings in as many kilometres. The structural calculations were carried out by Leon Moisseiff, using for the first time the "deflection" theory, developed in the 1880s by the Austrian engineer Joseph Melan, which demonstrated that excessively deep trusses were unnecessary and that stability under dead load and live load could be maintained with much more slender and flexible decks and towers. David Steinman, who translated Melan's work on his deflection theory into English in 1913, described the Manhattan Bridge as the first "to exemplify modern suspension bridge techniques".

Among several medium-span suspension bridges designed by Steinman himself, the 347m (1,114ft) span Florianapolis Bridge in Brazil (1926) stands out for several reasons. Unusually, it was an eyebar chain design, but, unlike the previous longest eyebar chain span, the 290m (951ft) Elizabeth Bridge over the Danube in Budapest (1903), it was erected by a new falsework-free method. It also marked the first use of *rocker* towers as a further development in flexibility from those of the Manhattan Bridge; and the stiffening truss was of a new, more economical design. In a sense, here was a New World rebirth in high tension, carbon-steel terms of the spirit of Telford's Menai Bridge exactly a century later, though in Europe the eyebar chain had continued as a popular type.

A new development for medium spans was the self-anchoring eyebar chain, first used in a 184m (605ft) Rhine bridge at Cologne (1915), in which the chain terminated at the ends of the stiffening truss rather than going into ground anchorages, so that the truss took the tension like a bowstring girder.

The alternative to parallel strands for wire cable designs were twisted rope strands. These were prefabricated, prestressed, and then brought to the site to be hung parallel over the towers and clamped into ropes. Notable Steinman bridges of this type were the Waldo-Hancock Bridge, Maine, and the St. John's Bridge, Portland, Oregon, both completed in 1931 and exhibiting the designer's taste for slender pinnacled towers of slightly "Gothic" eccentricity.

Meanwhile, the record for suspension spans had moved modestly upward. In 1924 the Bear Mountain Bridge took the lead at 497m (1,632ft), but only two years later, the much more massive Camden Bridge over the Delaware at Philadelphia was completed. The span was 533m (1,750ft), but more remarkable was the 39m (128ft) wide deck, carrying trains as well as road traffic. Three years on, and the Ambassador Bridge in Detroit took over as the longest span of any type – at 564m (1,850ft), just surpassing the Quebec cantilever bridge.

THE GEORGE WASHINGTON BRIDGE, NEW YORK
USA

As Gustav Lindenthal's assistant (see pp.96-7), the Swiss, Othmar Amman, was thoroughly familiar with his long dream of a bridge across the Hudson River. In 1923, he submitted his own scheme for a site upstream from Lindenthal's. Located away from the existing rail termini, it reflected the automobile's eclipse of the railroad.

Nearly everything about the structure of the George Washington Bridge, begun in 1927 and opened in 1931, redefined the state-of-the-art of suspension bridge construction. It was virtually double the previous maximum unsupported span, although it had modest side-spans of only 186m (610ft) each – even those on the Brooklyn had been substantially longer at 283m (930ft). The four suspension cables contained, between them, 107,000 miles (172,000km) of wire.

In some ways, despite its size, the George Washington Bridge presented its builders with a comparatively easy task. The west bank anchorage was ideal, driven direct into hard rock, although the east bank required the excavation of a huge cavern, which was then filled with mass concrete in order to create a firm anchorage. For the towers, there was no need for a prolonged ordeal of caisson-sinking, as

Below **The George Washington Bridge was the greatest cable-spinning commission yet for the Roebling company. On the Manhattan Bridge (1908) they had spun 33 tons of cable a day. Twenty years later, on the four mighty cables of the George Washington, this had risen to 61 tons per day.**
Facing page **The naked steel frames of the towers are overwhelming at close quarters.**

there had been on the Brooklyn Bridge. The George Washington's east tower could be constructed on dry land, although the piers for the west tower still had to be founded well below water level. For this, two huge sheet steel cofferdams were used, driven to a maximum depth of 26m (85ft) and then pumped out so that construction could take place in the dry. Nevertheless, on one occasion one of them burst and flooded, at a cost of several lives.

In his design for the towers, Ammann was influenced by Lindenthal and by the example of Roebling's Brooklyn Bridge. His inclination was for an appearance of monumental masonry, and he originally intended that the frameworks – each complex truss structure contained 20,000 tons of riveted steel – should be only the reinforcement for mass concrete, which would finally be covered with granite cladding. However, a growing general enthusiasm for the appearance of his unadorned steel skeletons coincided with cost restrictions brought about by the 1929 Wall Street Crash, and the client – the newly-formed New York Port Authority – decided to leave them as they were. Thus, the George Washington Bridge towers have remained ever since, a unique appearance achieved virtually by accident.

Originally, it was intended that the bridge have two decks, with eight lanes of roadway above and an urban rapid-transit rail line below, but during construction the latter was abandoned. Influenced by the deflection theory, which led to the conclusion that the longer and heavier a suspension bridge was, the less it needed a stiffening truss for stability, Amman calculated that the length of his bridge and its mass, totalling 56,000 tons of steel, were sufficient for him to omit such a truss altogether and rely on the 3m (10ft) thick plate-girder deck for sufficient stiffness. He was right; his bridge proved stable for the next 30 years. The second deck that it had originally been designed to accommodate was added in 1962 to allow for increasing traffic. Ammann, now 83, was guest of honour at the second inauguration; and in the course of the ceremony, a bust of him was installed on the bridge.

THE GEORGE WASHINGTON BRIDGE	FACTS
suspended span	1067m/3,500ft
depth of stiffening truss	8.8m/29ft
height of towers	183m/600ft
width of deck	36.3m/119ft
original depth to span ratio	1:350
depth to span ratio with stiffening truss	1:90

THE GOLDEN GATE BRIDGE, SAN FRANCISCO

USA

There was no better opportunity for surpassing the George Washington Bridge than on the West Coast, where gale-whipped ocean currents surged through the often fog-blanketed mile of the Golden Gate between San Francisco and the Marin Peninsula.

Long discussed, the idea was not taken seriously until 1917, when Joseph Strauss produced a design for a vast and rigid cantilever/suspension hybrid but during the ensuing decade of administrative wrangles, the George Washington demonstrated the pre-eminence of suspension structures for the longest spans.

Although Strauss remained in charge, a new design was drawn up, principally by his chief assistant, Charles Ellis. The weather and the risk of earthquake necessitated rigorous structural safety; the heavy shipping traffic demanded a deck of unprecedented height for clearance; and the glorious site invited a design of corresponding beauty. The towers were elegant pairs of vertical shafts, stepped back three times and linked at these points as well as at the top by deep horizontal beams. Both step-height and

Above **The Golden Gate's twin cables were barely thicker than the George Washington Bridge's four. Each 7,125 ton cable was made up of over 27,000 wires, spun back and forth across the gulf by continually travelling wheels, 24 wires at a time. Each strand, made up of 450 wires, was clamped into anchorages at each end and 61 strands were compacted into each cable, which was nearly 1m (3.3ft) thick. The wire rope suspenders were hung at 15m (50ft) intervals.**

beam-depth diminish as the towers rise, enhancing the bridge's grace. The suspended deck, 213m (700ft) longer but much narrower than the George Washington's, needed a stiffening truss, but it was shallower in relation to the great span than any yet achieved. The unforeseen rippling effect on the deck caused in later years by winds of much less intensity necessitated the addition of 4,700 tons of lateral bracing beneath, along its entire length.

Work began in January, 1933 on the shore anchorages, each with 32m (105ft) steel rods buried in over 50,000 tons of concrete. The north pier was built relatively easily on a bedrock ledge only 6m (19½ ft) below the water, in a cofferdam against the Marin cliff, but on the San Francisco side, the best option for Strauss's team was 335m (1,100ft) from shore, in water 30m (100 ft) deep—virtually the open ocean, amid frequent shipping.

The first idea was to found an oval fender 20m (65ft) down on bedrock, then excavate for the pier itself to a depth of 30.5m (100ft). As the sea was too rough for barges, a trestle was built out from the shore, but having twice been badly damaged, its deck was raised and its structure anchored more securely. The experience persuaded Strauss to drive the fender itself to the full depth, before the steel fender frame was placed in sections and concrete poured layer by layer. Strauss had then intended to float a huge caisson into the open fender, sink it with concrete and build the pier on top, but after a disastrous storm he decided to abandon it. He turned the fender into a cofferdam and pumped concrete into it, to the equivalent of six storeys depth, and after removing the remaining water, built the pier dry.

For flexibility and strength, the towers were riveted assemblies of hollow steel cells, just over 1m

(3¼ ft) square and between 6.9 and 13.7m (22½ft and 45ft) tall; due to the delays on the south pier, the Marin tower was built first. By August, 1935, cable-spinning began. The following fall saw the deck sections built out from both sides of each tower simultaneously, in order to balance the weight on the cables. On 27 May, 1937, pedestrians – 200,000 of them – were allowed on to the completed bridge, and on the morning of the following day it was opened to traffic.

THE GOLDEN GATE BRIDGE FACTS

suspended span	1,280m/4,200ft
depth of stiffening truss	7.6m/25ft
height of towers	227.4m/746ft
width to span ratio	1:47
width of deck	27.4m/90ft
depth to span ratio	1:168

Below **The Golden Gate Bridge has transcended its function as a bridge and a landmark to become both a local and a national ikon. The architect Irving Morrow was responsible for the red-lead colour and the unique detailing that make it, in a sense, the world's largest Art Deco sculpture.**

Right **This cutaway drawing of the monolithic south (San Francisco) pier shows the reinforcing frame in place on the concrete base around the perimeter of the pier (which encloses an area the size of a football field). The walls have been built up around the frame, the central pier has also been built up and the tower above completed; and the water, which had been pumped out until the tower was finished, has now been allowed to flow back in.**

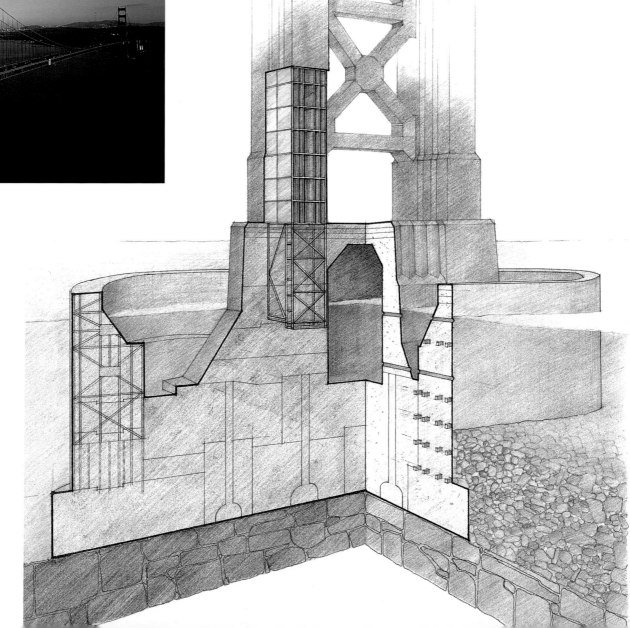

"GALLOPING GERTIE", WASHINGTON STATE
USA

Exactly contemporary with the Golden Gate Bridge, and in its way an even greater undertaking, the San Francisco Bay Bridge was a complex of twin 704m (2,310ft) suspension spans with an enormous common central anchorage, a tunnel and a 427m (1,400ft) cantilever, as well as extensive approaches. It was also a double-deck structure with a 9m (30ft) deep truss, which gave a depth/span ratio similar to those on earlier bridges, including the Brooklyn Bridge. By contrast, the Golden Gate's 7.6m (25ft) truss gave the high ratio of 1:168 – a trend continued by the next large American suspension bridge, which was yet another major structure for New York, the Bronx-Whitestone.

Designed by Othmar Amman, this austerely beautiful bridge, with its plain steel towers linked by a single deep cross-beam at the top, adopted plate-girder stiffening in the manner of the George Washington Bridge, but for a much lighter structure: its central span was 701m (2,300ft), giving a depth/span ratio of 1:209. Amman's great rival, Steinman, so dissimilar in his approach to tower design, was of a like mind when it came to the pursuit of slenderness and grace. His Deer Isle Bridge, completed in the same year, 1939, spanned 329m (1,080ft) with a deck only 7.6m (25ft) wide and 2m (6½ft) thick.

Below **The first European example of a plate girder suspension bridge was over the Weser at Rodenkirchen, designed by Fritz Leonhardt in 1938. The span was far less than on contemporary American examples – 378m (1240ft) – and the plate girder a little deeper, at 3.3m (11ft). It was destroyed in 1944 by bombs, rather than the torsional oscillation that caused Galloping Gertie's collapse, and it was rebuilt in virtually its original form.**

However, it was Leon Moisseiff who designed the ultimate in slender suspension bridges – and one which within four months of its opening in July, 1940, became the most notorious. The Tacoma Narrows Bridge over Puget Sound in Washington State had a suspended span that was 152m (500ft) longer than the Bronx-Whitestone's, but, as the builders anticipated a low volume of traffic, it was only a little over half the width – two carriageways plus walkways totalling a mere 11.9m (39ft). The plate girder was just 2.4m (8ft) thick, resulting in a depth/span ratio of 1:350, exactly equal to that of the original George Washington Bridge, but in a structure weighing only a small fraction of that mighty predecessor.

The bridge was promptly nicknamed "Galloping Gertie" because of its motion under quite light winds, which not only caused it to sway from side to side but also sent rippling waves along the deck. Attempts were made to dampen the movement with additional stays, but without success. On 7 November, 1940, a moderate wind of around 42mph (68kph) set up severe lateral twisting of the deck as well as longitudinal rippling. As at the "Hindenburg" disaster, three years earlier, there was a cameraman present to film Galloping Gertie's death-dance – hundreds of tons of writhing steel tearing themselves loose from their anchorages and plunging to destruction, a collapse which strikingly echoed descriptions of the loss of Ellet's Wheeling Bridge 86 years before.

The Tacoma Narrows Bridge was designed to be flexible, and it was not simply destroyed by sidesway or vertical deflections, neither of which had even approached their planned limits. As it turned out, all the major suspension bridge designers of the 20th century had ignored the lessons of history in their zeal for slenderness. Just like the Wheeling Bridge, the bridge at Tacoma Narrows was aerodynamically unstable. Subsequent wind tunnel tests showed that the narrow but solid plate girder presented a profile to the wind which encouraged oscillation, unlike an open truss, which fragmented the wind mass into smaller eddies and thus did not move nearly as much. At a certain intensity oscillation is torsional, and tends to increase in amplitude under its own momentum until the structure fails.

The consequences were immediate and far-reaching. The Bronx-Whitestone Bridge, already subject to noticeable oscillations, had a truss added, in this case above the level of the deck. And for the next quarter of a century, no major suspension bridge was built without a stiffening truss.

Seconds before the disaster, on 7 November 1940, ''Galloping Gertie'''s deck clearly exhibits the fatal torsional, or twisting, motion that tore it to pieces.

Right **The bridge dramatically collapses, revealing even more clearly its remarkable insubstantiality – the slender towers and narrow thin deck.**

''GALLOPING GERTIE''	FACTS
suspended span	853m/2,800ft
original deck width	11.9m/39ft
original deck depth	2.4m/8ft
rebuilt deck width	18.3m/60ft
rebuilt truss depth	10m/33ft
original depth:span ratio	1:350
rebuilt depth:span ratio	1:85

THE MACKINAC STRAITS BRIDGE, MICHIGAN
USA

At the time of Galloping Gerties's collapse, another and much larger plate girder design by Leon Moisseiff was at the active planning stage. This tremendous scheme was to cross the four-mile (6.4km) Straits of Mackinac, which link the Great Lakes Huron and Michigan. The Second World War intervened, however, and although Moisseiff was not subjected to the sort of public humiliation which Sir Thomas Bouch received after the destruction of his Tay Bridge 70 years earlier (see pp.70-71), Moisseiff's plate girder scheme for the Mackinac Straits had nevertheless been dropped by the time the project resurfaced, a decade after its inception. This time, Othmar Ammann, David Steinman and another engineer named Glenn Woodruff were appointed to report on how to bridge the Straits. The two great rivals disagreed; Ammann withdrew, and "Big Mac" became virtually Steinman's last, and without doubt his greatest, bridge.

The lessons of Galloping Gertie had been well learned. The Mackinac Straits are exceptionally hazardous – they are prone to frequent and sometimes violent gales, and are often blocked by thick ice-drifts in winter – and Steinman met these environmental challenges with a comprehensive armoury. Firstly, he ensured the stability of his cellular steel towers by constructing unprecedentedly massive pier foundations. They were made of steel and concrete totalling nearly a million tons, they were anchored in rock more than 60m (200ft) below water level, and they were built to withstand 51 tons per foot of ice pressure. Secondly, he designed for total aerodynamic stability of the span, with the 14.9m (48ft) roadway supported from beneath by a truss 11.6m (38ft) deep and 20.9m (68ft) wide, giving a depth/span ratio of exactly 1:100, the resulting open spaces between deck and truss further spoiling wind forces. Finally, his twin 620mm (24in) cables were designed with a huge margin of safety, which made them capable of carrying ten times the heaviest-conceivable moving load on the bridge. Each of the cables contains 37 strands comprising 340 wires each, each wire being about 5mm (⅕in) thick. In total over 69,000km (42,800 miles) of wire were spun for the cables.

The suspended span, at 1,158m (3,800ft), was second only to the Golden Gate when the Mackinac Straits Bridge was completed in 1957; and because of its exceptionally extensive side-spans – each 549m (1,800ft) – it is still, 36 years later, the world's longest suspension bridge overall. The total length from anchorage to anchorage is 2,626m (8,614ft).

Below **The three deep cross-beams on the towers above deck-level, although reminiscent of the four beams at the** Golden Gate (see pp.104-105), **have ornamental cutouts which are a characteristic feature of Steinman's work.**

Right **The unusual protrusion of the stiffening truss by some 3m (10ft) beyond the**

MACKINAC STRAITS BRIDGE	FACTS
constructed	1954-57
suspended span	1,158m/ 3,800ft
height of towers	168m/552ft
weight of structural steel	55,000 tons
weight of cables	11,000 tons

deck width on both sides of the Mackinac is one of the features designed to break up the force of high winds hitting the sides of the bridge. If anything, this response to the Galloping Gertie disaster is over-compensatory.

VERRAZANO NARROWS BRIDGE, NEW YORK

USA

Fittingly, Othmar Ammann's career, like that of his great rival David Steinman (see p. 108), reached its climax with his grandest concept. The Verrazano Narrows Bridge straddles the portal of New York Harbour, the focus of so many separate waterways themselves already bridged by numerous, varied, and historically important structures.

The bridge gathered most of the records: it had the longest span, just exceeding the Golden Gate's; its towers were second only to its Californian predecessor's; and it was by far the heaviest structure, with a greater weight of structural steel than any other bridge. It was the biggest bridge in history, but it was not notably innovative. Rather, it was a concentration of all the monumental simplicity and security towards which Amman's earlier designs had been leading. The towers, for example, are essentially a larger reworking of those on the Bronx-Whitestone Bridge, with a single deep beam at the head, and their simplicity is a stark contrast to the decorative style that Steinman was still using when he designed "Big Mac" – although even this was more restrained than some of Steinman's earlier bridges.

The deck structure was carried by four 0.9m (36in) cables – exactly the same size as those used on the George Washington Bridge 30 years earlier, although the increased length meant that the Verrazano Narrows needed 35,000 more miles (56,315km) of wire. The overall massiveness of the bridge was necessary in view of the huge volume of traffic that it was expected to carry: it was ultimately built with no less than 12 carriageways, on two decks over 30m (100ft) wide. The total depth, however, is only 7.3m (24ft), giving a depth/span ratio of 1:178. Visually, the effect is more like a single massive plate than two levels joined by a truss; and this, combined with the huge but simple towers, gives the whole bridge an extraordinary appearance of solidity and weight. Although it lacks the fascination of the George Washington, the sheer beauty of the Golden Gate, or the epic sweep of the Mackinac Straits, the Verrazano Narrows Bridge effectively marked the zenith and culmination of a whole century of American supremacy in suspension bridge design.

Right **The first gateway to New York Harbour from the Atlantic, the Verrazano Narrows Bridge is so huge that, even though the great depth of the stiffening truss houses a second lower road deck, the visual effect is more that of a single massive deep beam. The perspective of this photograph makes the cables seem slender, but in fact they contain enough wire to encircle the earth more than five times.**
Inset **Each of the cellular bridge towers contains 27,000 tons of steel. Once again the viewpoint alters the visual effect; looked at from above, the very deep beams at the heads of the towers can easily seem top-heavy and over-bearing, even for the mighty structure beneath.**

VERRAZANO NARROWS BRIDGE	FACTS
constructed	1959-64
suspended span	1,298m/4,260ft
depth: span ratio	1:178
clearance above water	66m/216ft
height of towers	207m/680ft
weight of steelwork	144,000 tons

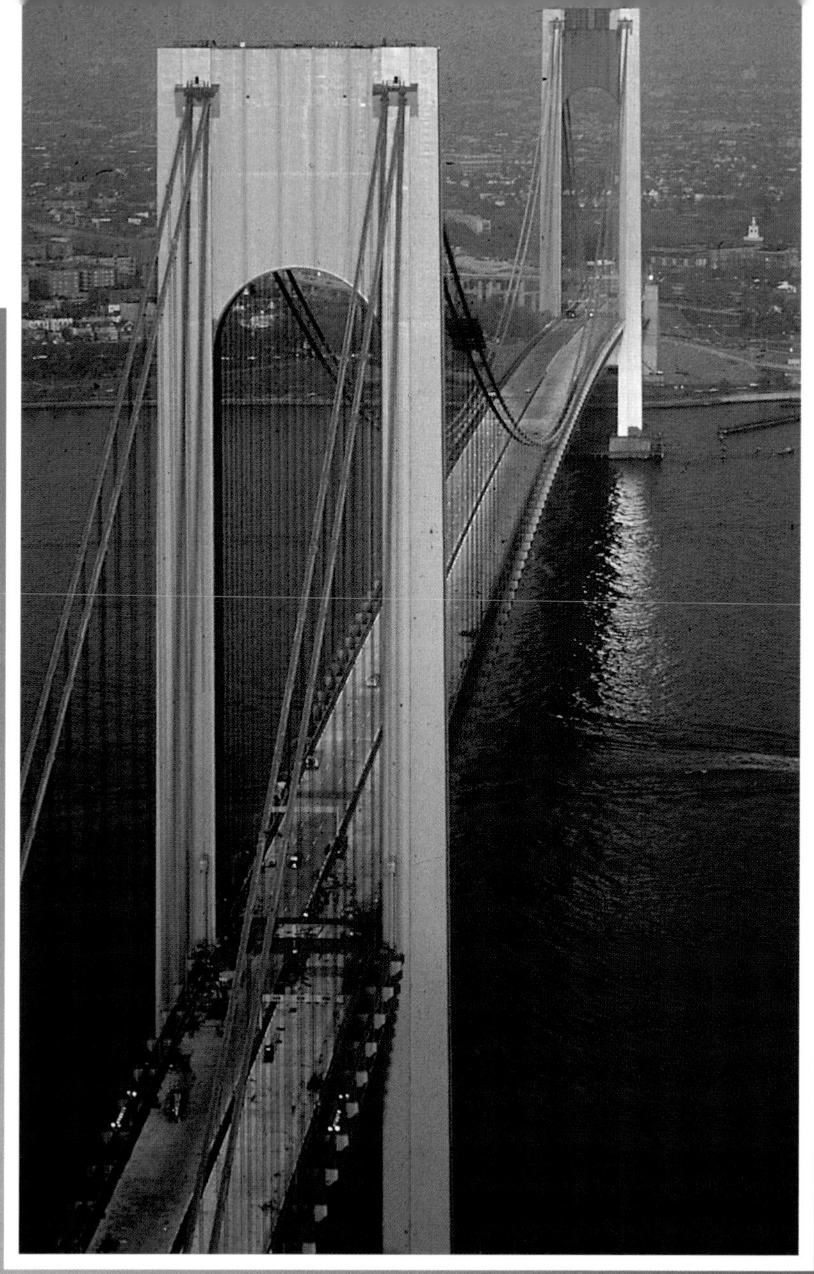

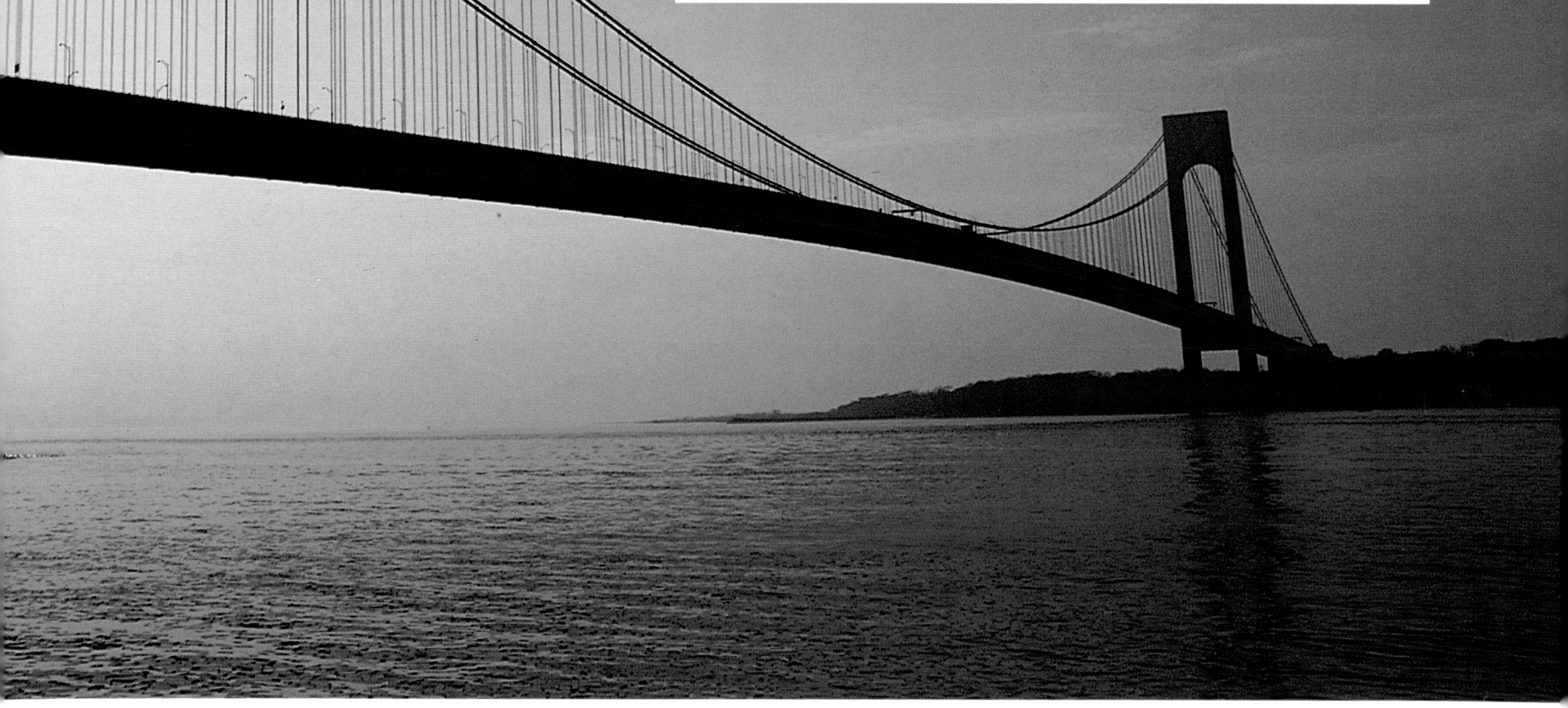

THE FORTH ROAD BRIDGE
SCOTLAND

The North American near-monopoly on long-span suspension bridge construction was finally broken in the early 1960s by three major European projects. Two of them broadly followed the design and construction practices perfected across the Atlantic, but the third broke spectacularly with precedent.

The need for a road link across the Firth of Forth in Scotland had been felt long before the first sod of earth was turned. Over three decades, as many different ideas had been examined and rejected: a two-span bridge alongside the Rail Bridge; the latter's conversion into a two-deck road and rail crossing; and a tunnel. Fortunately, in the final choice, economic and structural considerations coincided with aesthetic values: a single suspended span of 1,006m (3,300ft) with 408m (1,340ft) side-spans was built half a mile from the Rail Bridge.

Work began in 1958. The north tower was sited relatively easily on a submerged shallow ledge, but the south tower had to be sited on caissons sunk within a drained-out cofferdam. Though the British designers followed American convention with a trussed deck, 8.4m (27ft 6in) deep and 23.8m (78ft) wide, the whole bridge used far less steel than its American counterparts. With a main span three quarters the length of the Verrazano Narrows Bridge, for example, the Forth Road Bridge is almost three quarters lighter. The very slender legs were built in five large, vertical, welded box-sections, rather than small riveted cells, resulting in major savings in steelwork. The legs were joined, again unlike American examples, by even more delicate X-bracing. Indeed, the north tower, the first to rise to its full 156m (512ft) height, swayed considerably in light cross-winds until a damping system was introduced. Wire spinning began in late 1961, but severe gales that winter tangled them badly. Constant bad weather subsequently hampered construction so much that the bridge did not open until September, 1964.

Left **The first European suspension bridge to rival achievements in America, the Forth Road Bridge in Scotland represented a structural "half-way house" between the established transatlantic method and the new British approach. The steel truss used on American bridges is still present, but its more sparing and economical design is clearly evident here.**

Below **Completed 75 years later, the second Forth Bridge exemplifies wholly different structural concepts from the first and also demonstrates the growing supremacy of road over rail as a means of transport. The two Forth Bridges contrast and complement each other perfectly. A third crossing of the Firth of Forth is now under active consideration.**

THE TAGUS BRIDGE
PORTUGAL

In the late 1950s an international competition for the design and construction of a bridge across the estuary of the River Tagus near Portugal's capital, Lisbon, was won by an American consortium which included Steinman's design consultancy and in 1962 construction of the Tagus Bridge began. The intention was to make it possible to adapt the bridge at a later date to include a lower rail deck; and so the roadway was supported by a deep, 10.7m (35ft) truss, within which transverse X-stays were erected in pairs, one on each side, leaving a "hole" down the middle between them which was wide enough to carry twin rail lines. For maximum stiffness, the whole truss was designed to be continuous for the clear 2,277m (7,472ft) from south to north anchorage with sliding joints for expansion at the terminations and no fixed links with the towers. It was thus constructed as the world's longest continuous truss.

Below **This view of the Tagus Bridge clearly illustrates one of its most notable features – the exceptionally high deck clearance of 70m (230ft) required to allow the passage of very large vessels into and out of Lisbon.**

Not surprisingly, the bridge was thoroughly American in concept. The towers themselves were of riveted cellular construction, although they had light girder X-bracing somewhat like the Forth's, and they reached the great height of 190.5m (625ft), partly because of the 70m (230ft) clearance required by shipping. The Tagus Bridge also achieved a world record for the depth of the foundations on its south pier, where it was necessary to plunge 79m (260ft) below water level to the basalt bedrock; and for the first time in Europe, the stability and sinking of the caissons was controlled by compressed-air domes.

Opened in 1966, the bridge was originally known as the Salazar Bridge, but, after the dictator's downfall, it was renamed Ponte de 25 Abril (the country's Liberty Day). It remains mainland Europe's longest suspension bridge, with a main span of 1,013m (3,323ft) and 483m (1,586ft) side-spans.

Facing page **A light railway was originally envisaged to be installed within the truss beneath the deck, but in the mid-1990s a comprehensive refurbishment of the bridge was undertaken to add tracks for a much heavier, full-sized railway, as well as an extra road lane above. The towers and foundations were capable of handling the extra load, but the deck and truss had to be rebuilt and new cables spun. A short distance upstream is the 18km (11 mile) Vasco da Gama Bridge, a second Tagus crossing with a 420m (1,378ft) cable-stayed main span, built for Lisbon's 1998 Expo.**

THE SEVERN BRIDGE
ENGLAND/WALES

One of the unsuccessful entries for the 1960 Tagus Bridge competition was designed by Fritz Leonhardt (b.1909), who has become the doyen of European bridge engineers. After the Tacoma Narrows disaster, Leonhardt came to very different conclusions from the Americans on how to deal with aerodynamic forces on suspension bridges. Rather than use trusses to give maximum stiffness, he proposed that a narrow "aerofoil" deck should be used to minimize wind resistance and avoid the creation of the eddies that led to dangerous oscillation, and that it should be coupled with a single cable placed centrally over tall, narrow, A-shaped towers.

The English consultants Mott, Hay & Anderson and Freeman, Fox & Partners, were appointed in 1946 to design a new bridge over the estuary of the River Severn, England – before their appointment for the Forth Road Bridge, which they also designed. However, priority was eventually given to the Forth Bridge, the design of which was based upon the consultants' original proposals for the Severn. When they returned to the Severn project, in 1960, the engineer Gilbert Roberts of Freeman Fox carried out wind tunnel tests on experimental deck sections, and these led them to adopt an aerofoil-type section, similar to that envisaged by Leonhardt.

The deck design afforded savings in the weight of steel, and so did the design of the towers. The large boxes used on the Forth were a kind of half-way stage between the American riveted cellular type and the Severn, where the full side-widths of each tower leg were large steel plates, 16.6m (55ft) tall, bolted together on site. The legs had internal horizontal stiffening and were braced by deep steel portals, one immediately beneath the deck level for each tower and two above it. Although the Forth towers were much lighter than comparable American ones, those of the Severn, which were only about 30m (100ft) shorter, used just half its weight of steel.

Work began in 1961. The main problems in founding the piers were the Severn's swift flow and extremely deep tidal range. The west pier was particularly difficult, as the men were only able to excavate the site and place concrete blocks for the cofferdam during two 20-minute periods a day at low tide. After erection of the towers and the spinning of the main cables, the deck sections were prefabricated in 18m (60ft) lengths at a steelworks in Chepstow, floated down the river and lifted into position. The suspension system did include a major innovation – the hangers had an inclined zig-zag configuration which was a radical departure from the usual, 20th-century, American, vertical manner – and intended to improve structural damping.

The bridge was opened in September, 1966, and proved a triumphant vindication of the revolutionary deck design, which was subsequently adapted on several other large British-designed suspension bridges. Leonhardt himself has referred to it as the "first suspension bridge of the modern type". However, vehicle loadings increased greatly in the 1970s and comprehensive refurbishment became necessary after only 15 years. Between 1985 and 1991, all the hangers were replaced, the deck was repaired and strengthened, and the towers were reinforced by the insertion of tubular columns within the boxes.

Far right **The Severn Bridge, built to carry Britain's M4 motorway across the lower reaches of the country's longest river, establishing a vital link between South Wales and England. The innovatory zig-zag hangers are clearly visible. A second Severn crossing, with a cable-stayed main span, has now been completed.**

Left, top **Part of the Severn Bridge deck being lifted into position.**
Diagram **A section through the deck, which was 3m (10ft) deep at the centre, with each 9.9m (32ft 6in) half of the road surface cambering slightly downwards to a safety barrier; the remainder of the deck surface sloped steeply down to the lines of hangers at the perimeter, 22.9m (75ft) apart. Cantilevered outside this were 4.5m (15ft) footways, each designed to be extremely slender in section, in order to form the "trailing" edges of the aerofoil.**

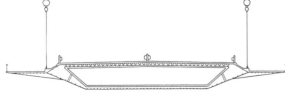

THE SEVERN BRIDGE FACTS
construction period 1961-66
suspended span 987.5m/3,240ft
side spans 305m/1,000ft
total height of towers 136m/445ft
deck clearance 47m/155ft
depth: span ratio 1:324

10

THE COMING OF CONCRETE

The invention of "concrete" undoubtedly long pre-dates the Romans' first use of it in around the 2nd century BC, but the Roman concrete containing lime and pozzalana – the volcanic powder which gave it both strength and waterproofing – virtually disappeared with the Empire. Although water-soluble lime mortar was used throughout the Middle Ages, knowledge of the Roman-type concrete virtually disappeared until the latter half of the 18th century, when, for example, John Smeaton developed new waterproof pozzolanic cements and used them to set the masonry of the Eddystone Lighthouse. Then, in the early 19th century, both natural cement and the new artificial "Portland" cement, which was invented and patented by Joseph Aspdin in 1824, became widely used for pointing and facing masonry, and for the mass concrete foundations in civil engineering works.

Experiments with "concrete" were also being carried out in France. In 1831 an architect named Lebrun put forward an unsuccessful proposal for a concrete bridge over the River Agoût. From about the middle of the 19th century, however, bridges in mass concrete began to be built in Europe and America and, as it was an "artificial stone", designs naturally followed the time-honoured shape of masonry bridges – the arch. The first major example in Britain was surprisingly late: the Glenfinnan Railway Viaduct in Invernessshire, Scotland, was constructed in 1898. Only the fact that the material had not been disguised with stone cladding or imitation voussoir markings distinguished it visually from hundreds of masonry prototypes.

The essential concept behind the reinforcing of concrete is the production of a composite material which combines the high tensile strength of steel (or originally wrought iron) with the compressive strength of mass concrete, the presence of metal rods negating the concrete's low strength in tension. In the first half of the 19th century, iron was occasionally embedded in concrete to strengthen it; Thomas Telford himself used iron ties in the Menai Bridge's concrete abutments, and later others experimented with floor members and even the hull of a boat. But the man generally credited with liberating the true structural potential of reinforced concrete was a French gardener named Joseph Monier. In 1867 Monier took out a patent for making plant tubs out of cement mortar strengthened with embedded iron netting. Other patents followed, first for railway sleepers, and then various building applications, including bridges: he built a 16m (52ft) arch in 1875.

In the United States, quite independently, a bolt manufacturer named William Ward also experimented with structures of iron buried in concrete; and a lawyer and engineer, Thaddeus Hyatt, became the first to analyze the stresses in concrete beams reinforced with iron. The Viennese Joseph Melan also brought scientific analysis to bear, developing an arch-based bridge design and building several examples in which a steel arch acted as centering for the poured concrete and then became its reinforcement when it hardened.

Ward and Hyatt both realized that in a beam of whatever section, gravity plus dead and live loading will ensure that the bottom part is in tension and the top in compression, and that therefore any reinforcement should be concentrated below, where it is most needed; and Hyatt seems to have been the first to use this principle to devise a beam with a T-shaped cross-section, having flat wrought-iron strips secured on edge lengthways beneath a concrete slab.

Credit for the practical development and first widespread use of reinforced concrete in bridges goes principally to François Hennebique in France and G. A. Wayss in Germany. In the 1880s Hennebique extensively researched the design of T-beams, substituting steel for wrought iron and bending up the ends of the reinforcement bars near the supports into the "compression zone", paralleling earlier work by Hyatt. Meanwhile, Wayss had bought the German rights to Monier's patent. By the mid-1890s his firm, Wayss & Freitag, had built numerous reinforced-concrete arch-bridges, some with spans exceeding 30m (100ft), but their designs were still relatively heavy, conservative imitations of masonry.

From Hyatt onwards, the proponents of reinforced concrete emphasized both its fire-resistant properties and the fact that it was virtually maintenance-free, but the latter was heavily dependent on

the quality of the design and workmanship, as it was crucial that the concrete should not "spall" or break off, leaving the exposed reinforcement to rust.

Hennebique took out patents in 1892 and, through his widespread network of contracting agents, ensured that his designs became structures. By 1900 there were over 3,000 of them, including about 100 bridges. They were the most forward-looking yet, and they had begun to exploit the material's potential for strength and slenderness. Hennebique's most notable bridge before the turn of the century was the Pont de Châtellerault (1899) over the River Vienne in France. It has two segmental-arch spans of 40m (131ft) flanking a similar central span of 50m (164ft). With rise to span ratios of about 1:10 their curve is shallow, but the most remarkable aspect is the thinness of the arches, which reduce to only about one hundredth of the span width at the crowns.

A year earlier, the Swiss engineer Robert Maillart had constructed his first bridge while still working as an assistant to Hennebique. This, the first reinforced-concrete bridge by the man many still consider to be the greatest artist in the field, was the modest Stauffacher arch over the Sihl River in Zurich. Although it is stone-clad and ornamented to a degree that makes it unrecognizable as a Maillart bridge, there is a slender three-hinged concrete arch underneath – a precursor of masterpieces to come.

Above **The 30m (100ft) span Zuoz Bridge in Engadin, Switzerland, 1901 – the first of Robert Maillart's bridges to proclaim openly its concrete structure (rather than being concealed in stone cladding).**

Previous page **Britain's first major reinforced concrete bridge, the Glenfinnan Railway Viaduct, consisting of 21 15.2m (50ft) arches.**

Left **Beneath the smooth concrete skin of the second Waterloo Bridge, built between 1939 and 1942, is a complexity of steel reinforcement (a small section of which is shown here), necessary for both strength and stability.**

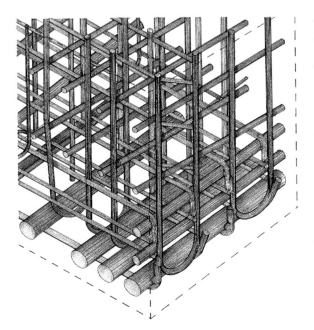

THE ART OF ROBERT MAILLART

SWITZERLAND

Like his first bridge, the Stauffacher, Robert Maillart's next two, at Zuoz and Billwil, were also 'prentice works. His break-through – almost literally – came with the 51m (167ft) span Rhine Bridge at Tavanasa in 1905. He had been concerned by the cracking which appeared after completion in the spandrel walls of the Zuoz Bridge, and at Tavanasa he omitted the areas that had cracked in the shape of triangular cutouts, their depth at the abutments reducing the extremities of the arch halves to slender fingers of concrete, on which the span seemed poised with unprecedented lightness and grace. Here for the first time was the mature combination of beauty, practicality and economy (Maillart's designs were usually cheaper than those of his rivals) which characterized his work for the remainder of his life.

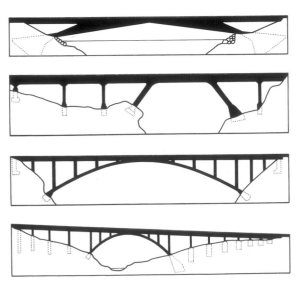

Facing page **The Salginatobel Bridge in the distance** (top left) **and from beneath** (bottom left). **Only a close-up view** (top right) **reveals the broadening of the structure at the piers for stability. The timber formwork for the arch** (bottom right) **was a major engineering feat in itself, with its crown 76m (250ft) above the valley floor.**

Left, top **The Tavanasa Bridge over the Rhine. As in many of Robert Maillart's earlier bridges, there is an odd and rather disconcerting contrast between the slender refinement of the bridge itself and the old-style solid masonry abutment.**

Bottom left **Just four examples of Maillart's seemingly inexhaustible design resourcefulness.**
1 **The Simme Bridge, Garstatt, 1939-1940, 32m (105ft)**
2 **The Eau-Noire Aqueduct, Châtelard, 1925, main span 30.4m (100ft)**
3 **The Schwandbach Bridge, 1933, 37.4m (123m)**
4 **Lancy-Genève project, designed 1936, main span 50m (164ft)**

In subsequent bridges, the varied conditions of site and purpose acted as a continual stimulus to his extraordinarily fertile imagination. On some, in place of the arrow-like cutouts, he removed the spandrel walls entirely, introducing instead slender vertical links between arch and deck – either rows of pencil-like columns or thin planes of concrete. Sometimes the line of the roadway would be emphasized with a solid parapet; at others it was lightened with open railings with the visual emphasis thrown onto the supporting structure. Again, as on the Eau-Noire Aqueduct near Châtelard, he designed a continuous box-girder beam instead of an arch, supporting it on thin splayed concrete legs.

Though by no means the climax of his career, the Salginatobel Bridge near Schiers, completed in 1930, has gained pre-eminence in Maillart's work on account of the spectacular grandeur of its site (spanning a precipitous gorge high up in the Alpine foothills of the Graubünden Canton) and the fact that it is his longest executed span – although it is still only a modest 90m (295ft). The entire bridge is canted upwards at a steady 1 in 34 inclination, meeting the rocky wall on the "short" side with spectacular abruptness and no visible abutment, so that the whole sublimely slender structure looks from the side as if it had been gently eased into its mountain fastness by a giant's hand. However, from the approaches to the bridge, it is its stability which is emphasized – the arch spreading at its footings to a width of 6m (19.7m) from the 3.5m (11ft 6in) wide deck.

In the ten years after Salginatobel, Maillart continued to explore new lines of development. He perfected structural treatments for skewed sections, oblique crossings and decks curved in plan; and in some of his last bridges he flattened the arches so much that the curve was barely perceptible – indeed in at least one instance he reduced the underside to two gently inclined straight lines meeting at the crown, as though the two halves of the bridge were lightly resting against each other.

At his death in 1940, Maillart's work was not particularly well-known outside his native Switzerland, but in the years that followed his reputation rose and he is now seen as a master of Modern art, architecture, and resourceful engineering. With the one notable exception of the Tavanasa Bridge, which sadly was destroyed by a landslide in 1927, almost all his bridges survive, the best of them exhibiting an eternally new-minted, breathtaking beauty. No 20th century bridge engineer has had more influence.

FREYSSINET AND THE PLOUGASTEL BRIDGE, BRITTANY
FRANCE

After Robert Maillart, Eugène Freyssinet (1879-1962) was the second great name in early 20th-century concrete bridge building. A graduate of the *École des Ponts et Chaussées*, he began his career spectacularly in 1907, while still a junior engineer at Moulins in south central France. The local highway department was preparing to replace three old suspension bridges over the River Allier, but when it designed such a costly stone successor for one that there was nothing left in the budget for the others, Freyssinet offered to build all three for the price of the one – to his own reinforced-concrete designs. Astonishingly, his offer was accepted, and he was given sole responsibility for the work.

The Veurdre, Boutiron and Châtel-de-Neuvre bridges, each with three spans of more than 70m

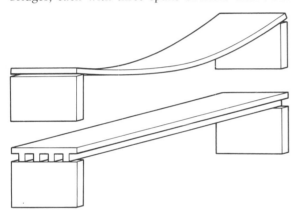

Below **Although small concrete bridges, with spans up to about 12m (40ft), could be made of unsupported reinforced slabs, longer slabs were liable to bow** (top). **However, until the advent of Freyssinet's prestressing technique, the addition of vertical webs beneath the slab** (bottom) **were the principal option for extending the possible span up to about 30m (100ft).**

(230ft), were duly erected within the estimate, and Freyssinet even found the resources to carry out a full-scale prototype test, in which he linked the bases of an arch with steel bars under tension to observe its pattern of movement. As a result, he was the first to discover concrete "creep" – the slow compaction which continues to take place after solidification. In anticipation of its effect on the Veurdre Bridge, he left an opening at the crown of each arch; and in 1911, a year after its completion, when the arches were observed to be sinking, he jacked their halves apart, raising the crown slightly, filled the openings with more concrete and thus fixed the spans permanently in a higher position.

The three bridges brought Freyssinet instant prominence and enabled him to set up as a designer. He constructed numerous other bridges, including a 100m (330ft) *un*-reinforced span at Villeneuve sur Lot. He also became involved in the development of reinforced-concrete buildings, as did Maillart in Switzerland. Unlike Maillart's buildings, however, some of Freyssinet's, such as the vast barrel-vaulted airship hangars at Orly, were on a spectacular scale; and when he came to tackle his greatest bridge, at virtually the same time as Maillart was working on the Salginatobel, it far outstripped the Swiss masterpiece in scale, if not in beauty.

The huge structure of three 180m (592ft) segmental arch spans across the Elorn estuary near the

Left **Casting the concrete arches of the Plougastel Bridge. A timber segmental arch (right in picture) was constructed with each end seated in a hollow floating concrete caisson. The caissons were firmly linked to stop the wooden arch deforming or collapsing. The first span of the bridge itself (left in picture) was cast on top of the wooden arch. When the concrete span had hardened the wooden arch with its caissons was eased free and positioned for the casting of the next.**

town of Plougastel in Brittany, which Freyssinet completed in 1930 after five years' work, broke all records for bridges in reinforced concrete. While the Salginatobel was designed for the light road traffic of its Alpine fastness, the Plougastel had to accommodate two decks, for rail above and road traffic below. His structure was of monumental clarity and simplicity: enormous, hollow, concrete-box arches rose to a height of 27.5m (90ft) at their centres, where they were 4.5m (14ft 8in) deep and 9.5m (31ft) wide. It was the first time that Freyssinet had employed a concrete-box structure, and he cast the three spans ingeniously by building a huge wooden arch which was supported at each end on a floating concrete barge and tied end to end for stability, and using it successively as formwork for each concrete arch. The twin trussed deck was then built on vertical fins of reinforced concrete, like some of Maillart's

but on a far more massive scale, between vertical box-sections of concrete at each end of the triple span.

The Plougastel Bridge's greatest importance lay in Freyssinet's investigation while he was working on it into the extent of concrete creep, which led to his realization of the practicability of prestressing. Although over 50 when the bridge was completed, Freyssinet was yet to establish the system with which his name is associated, and for which the construction of this masterpiece provided an essential key.

PLOUGASTEL BRIDGE FACTS

constructed	1925-30
length of spans	3 × 180m/592ft
height of arches	27.5m/90ft
width of deck	9m/30ft

THE BIRTH OF PRESTRESSING

Any building (or bridge) material can be "prestressed". A single rope across a river, stretched from one tree to another, has tensile stresses locked into it to reduce its flexibility. The principle of prestressing concrete, expressed at its simplest, is much the same: longitudinal steel strands in a concrete beam are stretched or tensioned and then anchored to the ends of the beam. This neutralizes the tensile forces created by dead, live and environmental loads – and thus the propensity of concrete for cracking – by the application of greater compressive forces. Far more effectively than with simple reinforcement, prestressing maximizes concrete's strength in compression and compensates for its weakness in tension.

An American engineer named P. A. Jackson was probably the first to formulate the idea of prestressing concrete when in 1872 he patented a system of passing iron tie rods through blocks and tightening them with nuts. Others experimented in subsequent decades, including Eugène Freyssinet who observed the action of creep for the first time when he used steel bars under tension to anchor the ends of his test arch for the Veurdre Bridge.

The technique of jacking the halves of an arch apart, which Freyssinet brought in as an expedient on the Veurdre Bridge to stave off disaster, was incorporated as an integral part of the design on two major bridges – the Candelier Bridge over the Sambre (1921), and the Saint Pierre du Vauvray (1923), at the time the world's longest concrete arch span at 131.8m (435ft). In both, he jacked the arches off the centering in two halves and inserted new concrete at the crown to withstand the effects of shrinkage. For Plougastel, however, he needed a precise

Below **The central span of the Medway Bridge in Kent, England, under construction in 1964. The box girder cantilevers are at their full 61m (200ft) extent and are awaiting the addition of the central supported section which is to be installed from the overhead linking gantry. The latest prestressing techniques allowed the vertical webs of concrete in the boxes to be only 230mm (9in) thick, although they do deepen to 10.7m (35ft) at the springings.**
Facing page, bottom **The completed Medway Bridge. Although visually unremarkable by today's standards, it was the biggest prestressed span of its day.**

evaluation of the extent of creep. He studied the phenomenon, got an answer, and concluded that prestressing with high strength steel bonded to the concrete was a viable structural system and in addition that it effectively created a new material.

At the age of 50, Freyssinet threw all his efforts and resources into establishing mass production facilities, but the venture collapsed through lack of an immediate market, and he was (temporarily) ruined. His salvation was another last-ditch rescue mission – this time halting the subsidence of major port facilities at Le Havre with prestressed foundations, which was a method which thereafter really took off.

After the Second World War, Freyssinet completed six single-span bridges over the River Marne – one of 55m (180ft) at Luzancy, which was built in 1945-46, and five of 73m (239ft) built between 1948 and 1950 at Esbly, Annet, Tribardou, Changis and Ussey. Although he later built longer prestressed-concrete spans, these exemplify his method. Their overall form was neither an arch nor a simple beam, but a shallow portal frame – a box of precast I-section girders reducing in depth towards the centre and cantilevered from the uprights of the portal (the *legs*) at the abutments. Vertical prestressing was introduced in the webs of the I-sections by casting the thin concrete around loops of high-tensile steel wire anchored in the upper and lower horizontal flanges, which were jacked apart during casting so that prestress would be introduced by release of the jacks after the concrete had hardened. Complete "voussoir" sections of the bridge were then hoisted into place and prestressed horizontally by steel cables,

which were passed through ducts and tensioned. Finally, to compensate for loss of strength through creep and shrinkage, Freyssinet placed jacks at the *hinges* (the bases of the portal legs) so that further compression could be introduced when necessary.

Freyssinet's Marne bridges, with their astonishingly shallow 1m (3¼ft) depth at the crown, were the harbingers of a vast wave of prestressed-concrete bridge-building that continues today. Pre-stressing itself can be carried out in two different ways. In pre-tensioning, the concrete is cast around the steel cables with them already under tension; when it hardens it is released from the mould and the wires severed from their ties, the bond between cables and concrete ensuring that most of the pre-tensioning is retained. In post-tensioning, on the other hand, the concrete is poured with voids incorporated, so that after it has hardened the cables can be threaded through, stretched, and anchored. The former is more commonly used for the precast production of building elements in the factory, the latter for casting larger members *in situ*. Both are used in bridge building.

Alongside the evolution of these techniques, the composition of concrete itself has undergone steady development, as have the design of formwork systems (with their consequences for the final appearance of the concrete surface) and the methods for compacting it by vibration before setting.

The 1950s and '60s saw such a volume of pre-stressed bridge-building world-wide that many books would be needed to encapsulate it. A few notable examples must suffice. In 1956 the first Lake Pontchartrain Bridge was completed in Louisiana, an extraordinary feat of precasting and high-speed erection in which no less than 2,170 separate 17m (56ft) spans were assembled from reinforced concrete slabs each supported by seven pre-tensioned beams, and placed on hollow prestressed-concrete piles by a floating crane. Totalling 38.35km (23 miles), it is exceeded in the *Guinness Book of Records* by only one other "longest bridging" – the second Lake Pontchartrain Bridge, completed 13

Above **The shallowness and slenderness of Freyssinet's Esbly Bridge are still remarkable today, nearly half a century after it, and four identical companions, were built across the River Marne, a few kilometres east of Paris, France in 1948-50. The triangular openings over the abutments further emphasize the bridge's apparent insubstantiality.**

years later alongside it, and just 69m (228ft) longer.

The German engineer Ulrich Finsterwalder developed the free cantilevering method of building long-span prestressed beam bridges, most notably used on the Rhine Bridge near Bendorf, completed in 1965 with a main span of 209m (682ft). This surpassed the new bridge built in 1964 to carry the M2 motorway over the River Medway in Kent, England, which has a 152m (500ft) main span. Here, slender 27m (89ft) wall-like piers, tapering upwards from 3m (10ft) to only 1.8m (6ft) wide, supported box girder cantilevers built out in sections simultaneously on both sides, linked 95m (312ft) away shorewards to viaducts of shorter spans and over the channel to a central suspended section constructed from 30m (100ft) prefabricated prestressed beams.

India now boasts one of the longest precast, prestressed-concrete beam bridges in the world. The Ganga Bridge at Patna is the longest road bridge in Asia, with 45 121m (397ft) cantilevered main spans. Like many such bridges, it was constructed with a moving launch gantry but, as sometimes is the case, one side of the bridge was completed and opened before the other was built – the incomplete structure presenting a lop-sided and spindly appearance, with reinforcing rods sticking upwards from half of the pier. It was finally completed in 1982.

To concentrate on materials (in the case of this chapter reinforced and prestressed concrete) is to ignore the fact that the form of most of the bridges described here could have been achieved in steel – indeed, many concrete box-girder bridges resemble steel box-girders in overall shape, and vice versa. Arches may be of steel or concrete, and equally, in recent years decks and towers of some suspension bridges have been constructed in concrete. The same goes for the most "recent" bridge form, the cable-stayed, as the next chapter will show.

THE ART OF PRESTRESSING

Left **This double prestressed concrete bridge on the Leventina Highway in Switzerland exemplifies a design feature of many modern Swiss road bridges: a relatively narrow box girder deepens over the pier in a quasi-arch, with the deck above cantilevered to a greater or lesser extent.**

Above **The Kylesku Bridge, designed by Ove Arup and Partners, complements its spectacular site across the turbulent narrows of Caolas Cumhann in north-west Scotland. The narrow, curved prestressed box girder is cantilevered from the pairs of V-shaped, sloping legs.**

Near right **The twin prestressed arch bridges across the Moesa Torrent on the south side of the San Bernadino pass in Switzerland were designed by Christian Menn in the spirit of Robert Maillart, but their spans of 112m (341ft) were longer than Maillart ever achieved.**

Far right **The immense Kocher Viaduct Geislingen in Germany is 1,128m (3700ft) long, with slender prestressed concrete piers carrying the 31m (102ft) wide deck up to 185m (607ft) above the valley floor. The slim box girder spans up to 138m (453ft).**

11
CABLE-STAYED BRIDGES

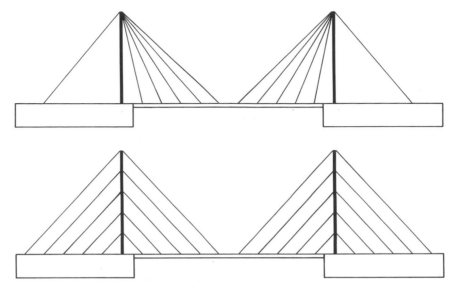

Right **The designs published in 1823 by the French engineer, Claude Navier, were extraordinarily similar in profile to many modern cable-stayed bridges, but they were never executed. In one design** (top) **five stays fanned out on the span side of the towers, and the** landward bracing was reduced to a single cable from the top of each tower without any horizontal stringers. In another (bottom), **Navier showed five stays on each side of the towers, this time parallel to each other at an angle of 40 degrees to the ground.**

Cable-stayed designs – in which bridge decks are directly connected to supporting masts by straight cables – burst on to the scene soon after World War Two. In 1952, Fritz Leonhardt designed a "family" of three cable-stayed bridges over the Rhine at Düsseldorf, the first of which, the Theodor Heuss or North Bridge, with main and side spans of 280m (919ft) and 108m (354ft), was completed in 1958. By then, the smaller Strömsund Bridge in Sweden, designed by another German, Franz Dischinger, had already been open for two years. Both were of steel.

No matter which is regarded as the first, the concept was hardly new. The idea of stabilizing and/or supporting a beam by ropes from a vertical support began with the booms, rigging, and masts of ancient Egyptian sailing ships. Some primitive bridges have decks stayed from above by ropes or vines; and the 16th-century Pons Ferreus illustrated by Verantius was as much a proto-stayed bridge as a suspension design.

Indeed, "cable-stayed" is not a long way from "cable-supported", and the history of suspension bridges is threaded with a less prominent, but still noticeable, stayed element. John Roebling's three major designs – Niagara, Cincinnati, and Brooklyn – incorporated networks of diagonal wire stays radiating directly from the tops of their towers to the decks, doubling up on the work of the main cables and their suspension wires. The stays helped both to support and stiffen the decks; and in the case of the Brooklyn Bridge, Roebling averred that "the floor, in connection with the stays, will support itself without the assistance of the cable".

Right **The Albert Bridge, Battersea, London, designed by R.M. Ordish and built between 1871 and 1873, has a main span of 122m (400ft). Diagonal stays of wrought iron radiating from its towers provide the principal support to the deck.**

Facing page, top **The Kniebrücke is the second of Fritz Leonhardt's family of three cable-stayed bridges across the Rhine at Düsseldorf. The main span is 320m (1,050ft) long.**

Facing page, bottom **The 61m-span (200ft) Maiden Castle footbridge at Durham, England, designed in steel by Ove Arup & Partners in 1974, exemplifies the elegance and clarity of form that cable-stayed designs can bring even to small bridges.**

Paradoxically, although Roebling regarded wire-staying only as a necessary back-up to the suspension system, he was probably the only designer before the advent of modern structural analysis who might have been capable of a successful, empirically designed stayed structure, but it seems that he never contemplated such a project. The first such design in modern times is attributed to a German named Löscher, who in 1784 published an illustrated account of a stayed bridge made entirely of timber – including the stays. In 1817, two British engineers, Redpath and Brown, built a small trussed footbridge stayed by wires from iron towers; and only a few years later two Frenchmen, Poyet and Navier, independently proposed designs remarkably prescient of present-day cable-stayed bridges.

In 1821 Poyet, who was an architect, illustrated tall towers and a trussed deck connected by six stays fanning out from both sides of the tops of the towers, with four levels of horizontal stringers connecting them. But it was not built, and adverse reactions followed the collapse of two contemporary chain-stayed bridges in England and Germany. The structural action of such bridges was simply not understood, and it was Navier himself who led future construction in the direction of suspension rather than stayed bridges.

During subsequent decades several intriguing, and perhaps unbuildable, pure stayed designs were published. However, those that were actually erected in Europe during the 19th century almost always combined radiating stays with catenary suspension cables like Roebling's. Among these was the Franz Joseph Bridge, opened in Prague in 1868, which most unusually supported the inclined stays from the suspension cable. Bridges designed by the French engineer Arnodin nearer the end of the century clearly divided the deck load between fan stays and the suspension cable. But the bridge which most anticipated modern designs was the small concrete Tempul Aqueduct, built over the Guadalete River in Spain by Torroja as early as 1925.

In 1938 Dischinger incorporated stay cables in a railroad suspension bridge near Hamburg, and in his investigations he realized that he could achieve stiffness and stability if the cables were made of high-strength steel wires under considerable stress. At the same time, Leonhardt was developing a concept for "orthotropic" steel decks – stiffened along the span length by longitudinal girders beneath the deck plates. After the War, Düsseldorf and Strömsund saw the first practical results of their work, which embodied all the virtues of safety, aesthetic quality, economy and simplicity. The way was open for a sweeping success story in cable-stayed designs, ranging from footbridges to all but the longest spans.

LAKE MARACAIBO BRIDGE

VENEZUELA

In 1957, the Venezuelan Government invited international tenders for a bridge to carry the Pan-American Highway across the 9km (5½ mile) northern neck of Venezuela's vast Lake Maracaibo. Of the 12 submissions, 11 were in structural steel; the exception was a reinforced and prestressed-concrete design, incorporating a 400m (1,300ft) cable-stayed navigation span, by the Italian engineer Riccardo Morandi. This was chosen for economic, aesthetic and political reasons: as well as its visual appeal, it was judged cheaper to maintain because of the toll that the near-Equatorial climate would be likely to impose on structural steel; it required the expenditure of less foreign exchange on imported materials; and it would provide valuable experience in prestressed-concrete construction for Venezuelan engineers and builders.

Although not quite the first cable-stayed bridge in concrete, Morandi's design was immediately seen as an epoch-making advance, and it established the powerful personal style which he was to reproduce with variations elsewhere, notably in Italy and Libya. As completed in 1962, the heroic structure rises steadily from the east along a viaduct of 109 spans to 45m (148ft) above the water, where it meets the central cable-stayed section of five navigation

Above **A motorist's eye view of the Lake Maracaibo Bridge, through the first of the pairs of cable stays.**

spans with side-spans at their outer extremities, and then descends along 19 spans to the western shore. The long eastern viaduct is of precast beams on plain vertical piers, but as the span-width and height increase towards the cable-stayed section, the piers change to V-shapes with progressively longer legs beneath, eventually creating a pinched H-shape. Under the main spans, the legs become double-X supports beneath the deck, with tall structurally separate A-shaped pylons on the perimeter carrying just single pairs of cable stays at their tips.

Early cable-stayed bridges like the Lake Maracaibo had few and relatively massive stays, requiring equally substantial anchorages in the deck, and the consequent long distances between stays meant that decks had to be correspondingly deep. From the late 1960s onwards, the trend was towards meshes of numerous, much finer supports, thus allowing reductions in deck-depth.

LAKE MARACAIBO BRIDGE FACTS

constructed	**1958-62**
total number of spans	**135**
height of pylons	**92.5m/303ft**
length of main spans	**5x235m/771ft**
width of deck	**14.22m/46½ft**

Left **The foreshortening effect of this photograph allows virtually the whole of the 8,678m (28,470ft) bridge to be viewed at a glance. This type of A-framed pylon, with A's in lateral pairs either side of the deck rather than singly straddling it, was a particular design "fingerprint" of Riccardo Morandi's cable-stayed bridges. The foreshortening also exaggerates the bridge's very gentle rise along its approaches to the 45m (148ft) deck clearance provided beneath the navigation spans.**

THE ANNACIS BRIDGE
BRITISH COLUMBIA

Steel cable-stayed bridges remain very common and are arguably the most suitable type for the very longest spans, and the Lake Maracaibo Bridge established a healthy breed of prestressed-concrete cable bridges which continues today, but there is a third and more recent option for the superstructure – composite construction, in which components of different materials act together as parts of the same structural unit.

Reinforced concrete can be regarded as a composite, although in practice "composite construction" usually means reinforced-concrete slabs plus steel girders. In a bridge, a thin reinforced- or prestressed-concrete deck will be supported by a frame of lateral and longitudinal girders.

The first cable-stayed bridge constructed in this way was the Annacis (now the Alex Fraser) Bridge, over the Fraser River near Vancouver, British Columbia, which was completed in 1986. The design of cable-stayed bridges was initially very much a

Below left The Annacis Bridge at the most precarious moment of its construction – the deck is built out from both sides of the towers as a balanced cantilever. Below right The 465m (1,526ft) main span of the Annacis Bridge safely joined to form a continuous whole. The side-spans are each 183m (600ft) long.

European speciality, and it took many years for the rest of the world, particularly North America, to catch up. The Annacis Bridge is doubly notable, therefore, not only for its pioneering composite deck, but for beating all contemporary rivals in length. Its twin concrete towers, 154.3m (506ft) high, are H-shaped with each column rising vertically to the deep cross-beam beneath the deck, then bending inwards to a second cross-beam and thereafter continuing vertically.

The deck is remarkably thin. The precast concrete panels are only 220mm (8.7in) thick, supported by a 2m (6.6ft) deep steel frame, which was built out from the towers as a balanced cantilever, with the slabs laid behind as it grew. Since such a slender deck requires continuous support, a cable is attached to the frame every 9m (30ft) throughout its length. The cable configuration is a semi-fan shape, and the cables closest to the towers are anchored about halfway up, just above the second deep transverse beam.

THE QUEEN ELIZABETH II BRIDGE, DARTFORD

ENGLAND

Left **Each of the main piers supporting the pylons of the Queen Elizabeth II Bridge was a concrete caisson 60m x 30m x 23m deep (197 x 98½ x 75½ft), precast in Holland and towed across the North Sea by tug. After being positioned over the site, they were sunk into place by flooding.**

Above right **Fabrication in progress of the steel girders for the deck, at the factory of Scott Lithgow Ltd in Scotland, to the design of Cleveland Structural Engineering.**

Facing page **The two halves of the deck, cantilevering outwards from the towers and approaching their meeting point in the summer of 1991. The contrast in weight and texture between the steel towers and the reinforced-concrete legs is the unfortunate consequence of having to satisfy in the design the conflicting constraints of the existing motorway and the requirement for shipping clearance.**

Large cable-stayed bridges came relatively late to Britain, as they did to North America. The first with a main span exceeding 1,000ft (305m) was Scotland's Erskine Bridge, completed in 1971. With single pylons, centrally mounted between the carriageways and supporting only single cables, it reduced the cable-stayed style to its simplest possible visual elements. Ten years later, the 240m (787ft) Kessock Bridge, also in Scotland, introduced the large harp configuration to the United Kingdom. And after another 10 years, the new Queen Elizabeth II Bridge from Dartford to Thurrock over the River Thames became for a few months Europe's longest cable-stayed bridge, with a central span of 450m (1,476ft). The bridge was designed to carry four lanes of road traffic, but, unusually, these all travel in the same direction, southwards, as the northbound flow on London's orbital M25 motorway runs through the two Dartford Tunnels.

The Queen Elizabeth II Bridge is a good example of how difficult local conditions can strongly influence a design, not necessarily to its aesthetic advantage. Here, the constraints of the 57.5m (188ft) navigation clearance, which was required for very large shipping, including the possibility of the Queen Elizabeth II liner, together with the conse-quent extremely high impact resistance to the caissons and piers, led to a design with steep approach viaducts from the M25 arriving at a slender composite steel and concrete deck – only 2m (6.6ft) deep – supported by pylons whose cross-section changes abruptly at deck level, from massive reinforced concrete rising 53m (174ft) from their piers to slender steel towers, towering a further 84m (276ft). The onerous conflict between the high navigation clearance required and the position and orientation of the existing motorway on both sides, led to a further, and most unusual compromise in the bridge's design: the arc of the central span is not symmetrical, but peaks closer to the north than to the south pylon. The diference is slight – indeed, it is a tribute to the bridge's designers that the compromise is so well concealed. The cable supports are of a semi-fan configuration, similar to those on the Annacis Bridge, and the design of the whole superstructure is largely the work of the German bridge engineer, Dr. Helmut Homberg, who crowned a 60-year career with this structure. However, for all that the Bridge makes a strong visual impact with its high-sweeping central span, it is difficult to avoid an impression of ungainliness and mismatch between the over-massive legs, the stark steel towers, and the narrow deck.

THE QUEEN ELIZABETH II BRIDGE FACTS

period of construction	**1988-91**
total cable-stayed length	**812m/2,664ft**
total approach viaducts	**2km/1¼ miles**
total height of pylons	**137m/450ft**

THE SKARNSUNDET BRIDGE
NORWAY

After the construction of the Annacis Bridge, several cable-stayed structures were completed with near-comparable spans, but its record was broken only in December, 1991, and then by two bridges – the 490m (1,608ft) Ikuchi Bridge in Japan, one small link in the vast complex joining the islands of Honshu and Shikoku (see p. 152) and the 530m (1,739ft) Skarnsundet Bridge in Norway.

The new Norwegian bridge replaces a ferry service across the Skarnsundet Strait in the Trondheims-fjord. Although it is an important link, the volume of traffic did not justify more than two carriageways, with the result that, for its length, the bridge is extraordinarily slender and narrow – only 2.15m (7ft) deep and 13m (42.6ft) wide – similar

dimensions to the Tacoma Narrows Bridge (see p. 106). To banish any further disastrous parallels with that ill-fated structure, the designers opted for a hollow, triangular, concrete box-section for the deck, which provided aerodynamic stability and stiffness under live loading; and to compensate further for the overall narrowness of the bridge, they chose A-shaped pylons, again in concrete, rising to a height of 152m (499ft) above the water. In line with current practice, the deck is supported by fans of numerous cables of galvanized wire. Beneath the landward extremities of the side-spans are concrete piers 27m (88ft) apart; and the hollow concrete box-sections of the deck supported by these are given extra mass as counter-weights by being filled with ballast.

Right **The original proposal for the Skarnsundet Strait was for a suspension bridge similar to others built to span comparable distances in Norway during the last 25 years. However, it is difficult to imagine even a modern suspension bridge fitting its beautiful site more elegantly than the cable-stayed structure which was finally adopted. Of the bridge's total 1,010m (3,314ft) span, just over half is suspended. In 1993 its record cable span was exceeded by the 602m (1,975ft) span Yangpu Bridge in Shanghai, and a year later by the Pont de Normandie (see pp. 150-151).**

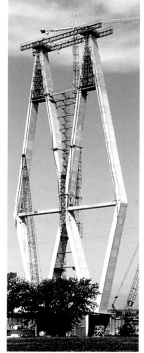

On the other side of the world, another quite different record for cable-stayed bridges has been set – and unlike the Skarnsundet with its longest span, it is one that will not be exceeded for a very long time, if ever. The 380m (1,250ft) main span of the Fred Hartman Bridge, near Houston, Texas is not re-

markable for cable-stayed structures near the end of the 20th century, but the width of it is. Completed in 1994, it replaced the Baytown Tunnel between the refinery centres, Baytown and La Porte. The volume of traffic here demands eight lanes, an expanse of deck totalling 47m (156ft) – virtually as wide as Sydney Harbour Bridge – in what is effectively a double bridge, supported by four planes of stay cables. Theoretically, it can carry 200,000 vehicles a day.

The pylons are unique double-diamonds in composite steel/concrete construction, the configuration acting like a huge truss to withstand anticipated hurricane wind forces much more effectively than H-shaped towers of the same width. An aesthetic spin-off from the shape is that the lateral width of the box sections forming the individual legs can be as narrow as 2.1m (7ft), tapering at the sides between 7.3m (24ft) at the base and 4.6m (15ft) at the tip.

Above left and centre **The double-diamond form of the towers makes the Fred Hartman Bridge in Texas unique among cable-stayed designs. The two roadways, which are not connected, and are each supported by two planes of cables, pass side by side through the openings of the double diamond. The photograph on the right shows one of the prestressed-concrete towers at its full height.**

VARIATIONS ON A THEME

The advent of cable-stayed bridges has been a liberating challenge, both aesthetically and structurally, to bridge designers in the latter half of the 20th century. They are relatively inexpensive to build, attractive and suitable for nearly every scale from small footbridges to very long spans, and they lend themselves to an almost infinite variety of shapes. The pylons can support a single span from one end or two spans from a central or eccentric position, as in Cologne's Severin Bridge (1960). They can be placed at each end of and supporting a central span with side-spans at the extremities; or there can be several of them supporting a multi-span bridge like India's Ganga Bridge (1969). The pylons can be single columns in the centre of the carriageway, with a single plane of cables supporting the bridge; they can be pairs, linked or not by cross-beams; or they can

Below right and facing page **Cable-stayed bridges lend themselves to a multitude of permutations. In the Khölbrand Bridge, Hamburg, Germany, 1974, the base of the A-frame tower curves in on itself to enclose the deck. The approach viaducts and box-girder deck are in concrete, as are the main piers, but the towers and central span are in steel.**

have A-shapes or other even more outlandish configurations. They can also lean backwards, as in the Alamillo Bridge at Seville's Expo '92, or forwards, as in Tasmania's Batman Bridge' (1968).

Decks can be concrete, steel or composite; and the cables themselves, ranging from one to many, can have harp, fan or hybrid configurations. The topmost cable can be attached to the tip of the mast or well down it, as in Tokyo's Arakawa Bridge (1970). The spread of cables over a side-span can equal that for the balancing half of the main span or it can be substantially less, as in Ove Arup & Partners' unbuilt design for the Second Severn Crossing (1990); and although the cables are almost always exposed, they can be enclosed, as in the concrete sleeves of Christian Menn's remarkable Ganter Bridge in Switzerland (1980) (pp.146-7).

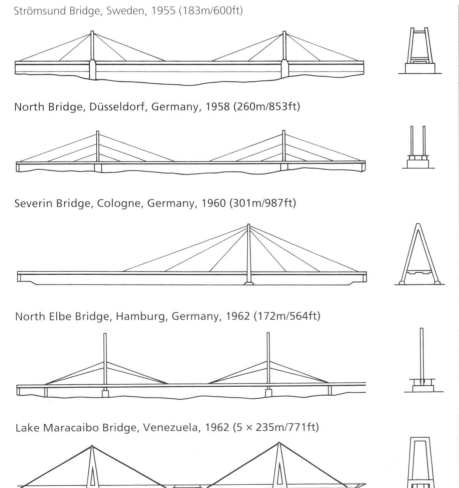

Strömsund Bridge, Sweden, 1955 (183m/600ft)

North Bridge, Düsseldorf, Germany, 1958 (260m/853ft)

Severin Bridge, Cologne, Germany, 1960 (301m/987ft)

North Elbe Bridge, Hamburg, Germany, 1962 (172m/564ft)

Lake Maracaibo Bridge, Venezuela, 1962 (5 × 235m/771ft)

Erskine Bridge, Scotland, 1971 (305m/1,000ft)

Bratislava Bridge, Slovakia, 1972 (303m/994ft)

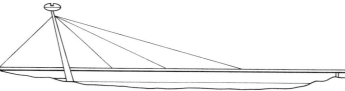

Mesopotamia Bridge, Argentina, 1972 (340m/1,116ft)

Speyer Bridge, Germany, 1974 (275m/962ft)

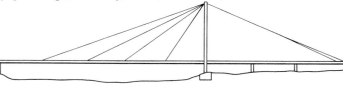

Kiev Bridge, Ukraine, 1976 (300m/984ft)

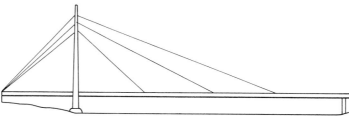

Barrios de Luna Bridge, Spain, 1983 (440m/1,444ft)

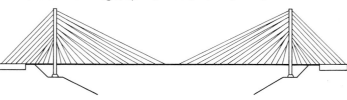

Sunshine Skyway Bridge, Florida, USA, 1986 (366m/1,200ft)

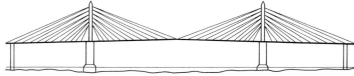

THE ART OF CABLE-STAYING

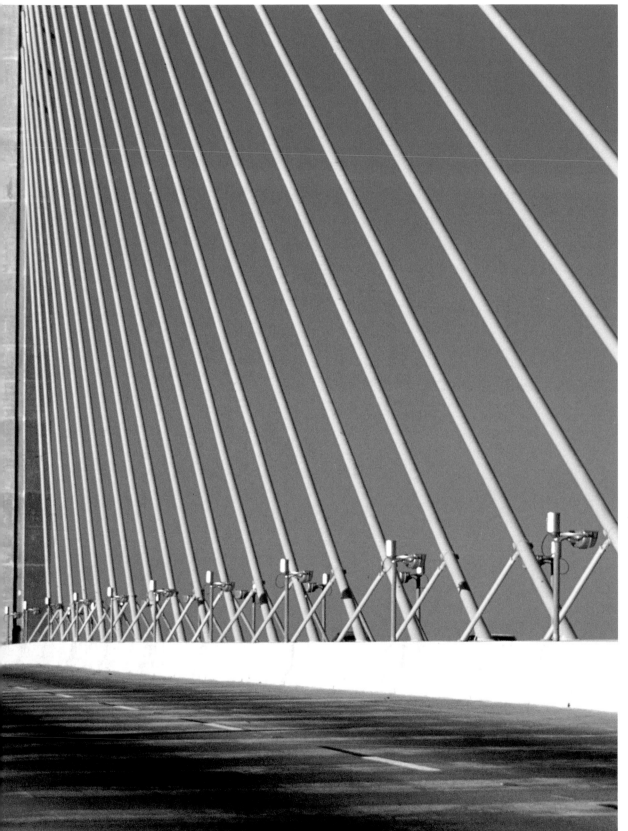

Main picture and facing page, top **The Sunshine Skyway Bridge over the Tampa Bay, Florida, USA. When an earlier bridge was struck by a freighter and badly damaged in 1980 the opportunity was taken to create an entirely new design. Completed in 1986, the replacement bridge extends for 6,670m (21,877ft), and bridges the central navigation channel of the Bay with the largest concrete span in North America. In addition, the great height of the deck was determined by the 53.4m (175ft) clearance necessary for shipping.**

Facing page, bottom **An even greater clearance of 61m (200ft) was required beneath the 404m (1,325ft) main span of the St. Nazaire Bridge in Brittany, France. Completed in 1974, it provides a marked contrast to the Sunshine Skyway Bridge. The chequered pattern on the inverted-V towers is certainly arresting, but the huge slab-like piers beneath have an oppressive quality that the American bridge wholly avoids.**

part four

The creation of attractive and practical bridges still poses a challenge to today's engineers. Inspired designers are drawing on new structural techniques and improved materials to extend the possibilities of existing bridge types and evolve wonderful new hybrid designs to take us into the third millennium.

The Barqueta Bridge at Expo '92, designed by J.J. Arenas and M.J. Pantaleon. Its hybrid arch/cable structure exemplifies the endless inventiveness of modern bridge designers.

12

THE STATE-OF-THE-ART

Although complex in detail, the history of bridge technology and use is essentially that of the developing relationship between available materials and structural possibilities under changing requirements.

Until the end of the 18th century, the masonry arch was the ideal combination of material and structure, both for durability and load, although the Grubenmanns demonstrated the more-than-equal potential of timber for long spans; and masonry and timber bridges have continued to be built into modern times. Compared with the all-conquering march of prestressed concrete, new stone bridges are a rarity, but the lofty and daringly slender stone arches on the facade of the Pabellon del Futuro at Expo '92 just might open up new possibilities, and as conservation-consciousness grows, restoration techniques for old masonry structures are improving rapidly.

Old timber bridges, which in the past had a high mortality rate, are now often lovingly preserved. In the United States, for example, some of the relatively few remaining covered bridges are cherished as local, and national, treasures. Not all is conservation of the old, however. The technique of lamination, in which layers of timber were glued together, often as an arch, used in some early 19th-century British railway bridges, is now being reintroduced in small-scale bridge-building using improved modern adhesives.

Materials, techniques, use – always the three interacting spheres. The Industrial Revolution brought railways, but at the same time it introduced cast iron (and later wrought iron and steel), enabling the great designers of the 19th century to exploit the strengths of the latter to meet the demands of the former. Spans were launched further and carried

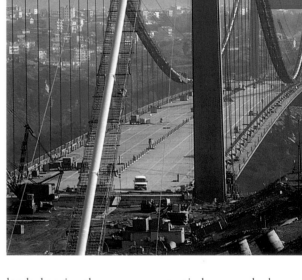

Right **The second Bosporus Bridge in Turkey, shown here shortly before its completion in 1988, represents the "state-of-the-art" in European suspension bridge design. Its main span of 1,090m (3,580ft) is slightly longer than that of its predecessor, and although it also has an aerodynamic deck, traditional verticals, replace the previous "zig-zag" cables.**

loads heavier than masonry or timber ever had or could. Cast iron, and subsequently steel, arches were built – and steel arches continue to be built.

In meeting the large and urgent need for road bridges of all lengths which has swept the world since the end of the Second World War, the leading edge of design development was first in steel and then in pre-stressed concrete. Segmental concrete box construction offers elegant solutions to the varied geometries of modern highways, and it is economic over both short and long spans. A new interaction between material and structure has recently been introduced on the Sylans and Glacières Viaducts in France, where the decks are supported by lightweight, cheap, triangular trusses in prestressed concrete. Although very long box-girder spans have been achieved without intermediate supports, some do use inclined struts to subdivide the length and thus reduce the

Left **The Bloukrans Bridge in South Africa, completed in 1983. This 272m (892ft) span prestressed-concrete arch crosses the Van Stadens Gorge, a spectacular stony chasm in the southern part of Cape Province. With its clarity of form it clearly echoes the designs of Maillart, but on a far larger scale than the Swiss master was ever enabled to build.**

necessary depth of the box – the outstanding example of this is the vividly red-painted Sfalassa Gorge Bridge in Italy.

The box-girder is equally adaptable for moveable bridges – the "state of the art" undoubtedly represented by West Seattle's Low-Level Swing Bridge. The 146m (480ft) centre span divides into two equal halves, which pivot on their support piers to lie parallel to the river banks when the bridge is opened. Each leaf is balanced by a 53m (173.5ft) tail span, so that in all some 15,000 tons of concrete glide apart and swing into place in about two minutes.

For spans beyond that for which an unsupported box is stable, cable-staying can be employed, bridging gulfs over which, only a few years ago, only suspended decks would have been considered suitable. Successive 300m (1,000ft) and 450m (1,500ft) "barriers" for cable-stayed spans were surpassed without mishap, and fears that France's Pont de Normandie would be "a bridge too far" were allayed when it opened in 1994 (see pp. 150-151). Aside from sheer size, the most aesthetically striking cable-stayed bridge to have been built in recent years is the Ganter Bridge in Switzerland (see pp. 146-7). It hides its cables in concrete sheaths, and in the United States the structural implications of this have been followed through in the first "fin-back" bridges, like the one completed at Barton Creek, near Austin, Texas, in 1987. Here, support to the balanced-cantilever side and mainspans is provided by a thin vertical web of prestressed concrete –

Above **In Vancouver, the Skytrain, the city's new urban lightweight rail system has been provided with the first large cable-stayed bridge to be built exclusively for this resurgent mode of transport. The 350m (1,115ft) span structure was opened in 1990 across the Fraser River. Nearby is the Pattuilo Bridge, built 1937-8.**

a solid "fin" above the deck level.

Apart from the brief ascendancy of the steel cantilever between 1889 and 1929, suspension bridges have been "state of the art" for the longest spans for nearly 200 years, and they look set to continue in that role, despite their potential aerodynamic instability. The Severn Bridge has had several successors adopting similar aerofoil deck configurations, including the Humber and Denmark's East Bridge.

Large suspension bridges have also become beneficiaries of the conservationists. After a design competition for a new Williamsburg Bridge in 1988, the 85-year-old structure was reprieved and rehabilitated; and a 1990 study found that the Golden Gate Bridge was capable of accommodating a second, rapid transit deck beneath the existing roadway.

Left **For sheer monumentality in steel box-girder bridge construction the Ponte Presidente Costa e Silva Bridge across the Guanabara Bay between Rio de Janeiro and Niteroi in Brazil is still unequalled since its completion in 1974. Totalling some 10km (6 miles) in length, the three main spans alone are of 200m, 300m and 200m (656ft, 984ft, 656ft).**

THE GLADESVILLE BRIDGE, SYDNEY
AUSTRALIA

The Gladesville Bridge across the Parramatta River, quite close to Sydney Harbour Bridge, was a milestone in prestressed-concrete arch construction when it was completed in 1964; indeed the only structure of the same type to have exceeded it in span is the Krk Island Bridge in the Adriatic. But the Gladesville embodies a curious paradox. Unlike many of its type built this century, it is segmental and was deliberately constructed on exactly the same principle of voussoir units as a Roman arch bridge – the difference being that the Gladesville's units were large, hollow and precast, rather than small, solid and hewn.

The Gladesville Bridge was originally tendered in 1957 as a steel cantilever, but an alternative concrete-arch proposal by a contractor was considered, developed and subsequently submitted for a report to the great Eugène Freyssinet (see pp. 122-5), whose recommendations were accepted. The final design was by the firm of English consultants Maunsell & Partners.

Temporary steel framework was erected across the River, with deep trusses leaving a 61m (200ft) opening for navigation. Each day five or six voussoir units, up to 6.1m by 6.9m (20ft by 22ft 7$\frac{1}{2}$in), were hoisted from barges into position and concreted together, and the complete arch was then jacked away from the formwork in a similar manner to that developed by Freyssinet 40 years earlier.

The deck, also made of precast, prestressed concrete, is a diaphragm structure designed to be as light as possible, as it has to bear a series of point loads on the arch. The deck itself is a much shallower arch, supported in 30m (100ft) long units above the main structural span by pairs of prestressed-concrete piers topped by deep, deck-wide beams immediately beneath the deck structure.

Below Despite the relatively flat land on either side of the Gladesville Bridge, and the need for clearance for shipping, its designers have achieved a fine compromise. The length and curvature of the arc of the roadway allow a very gentle rise for drivers, and form a balanced and graceful union with the main arch of the bridge.

THE GLADESVILLE BRIDGE	FACTS
constructed	**1959-64**
arch clearance	**40.7m/134ft**
arch span	**305m/1,000ft**
total length	**488m/1,600ft**
rise to span ratio	**1:7.5**

NEW RIVER GORGE BRIDGE, WEST VIRGINIA
USA

After 45 years as the world's longest steel arch, New York's Bayonne Bridge was finally exceeded in 1978. The New River Gorge Bridge, built to shorten the main north-south route in this part of West Virginia by some 64km (40 miles), is 15m (50ft) longer than the Bayonne; and it is almost the world's highest as well. This height was the factor that ruled out multi-span truss construction, while a suspension solution, which would otherwise have been perfectly feasible, was eliminated because of the potential hazard to aircraft from the towers. As the gorge was far too wide even for a "state-of-the-art", prestressed-concrete arch, the only realistic answer for a bridge of this height, scale, location and remoteness was a single span of high strength steel (which also weathers without corroding and thus avoids any need for regular painting).

The New River Gorge Bridge was designed by Michael Baker Jr. Inc. of the West Virginia Department of Highways. Concrete footings were built into the Gorge side for both the arch and the steel piers that would carry the deck, and then the piers and

Below **It is difficult at first to appreciate the scale of the New River Gorge Bridge. The single toy-like vehicle visible on the right gives some indication of its size, as does the fact that the vertical truss supports are 42.5m (139ft) apart – nearly 1½ times the entire span of its ultimate progenitor, the Ironbridge at Coalbrookdale.**

deck, which were all truss assemblies to reduce lateral wind loads, were built out to points above each arch footing. After this, each half of the arch was erected in sections, tied back by cables for stability. The major construction innovation came in joining the two arch halves together. Previous long-span steel arches, like those at Bayonne and Sydney Harbour, had been jacked apart slightly when the two halves were virtually touching, until the stresses in the arch reached their design value, and then steelwork had been placed to complete the arch. At New River Gorge, the designers dispensed with this, relying instead on computer calculations of the stresses and accurate fabrication of the structural pieces.

NEW RIVER GORGE BRIDGE	FACTS
constructed	**1975-8**
height of deck	**267m (876ft)**
arch span	**518m (1,700ft)**
total length	**924m (3,030ft)**
width of deck	**22m (72ft)**

THE GANTER BRIDGE
SWITZERLAND

Other bridges may be longer overall, of greater span, and even taller, but of the hundreds of slender, striking structures built after the Second World War to carry new roads around the mountains and across the valleys of central Europe, none offers a more arresting profile than the remarkable Ganter Bridge on the new Simplon Pass road above the town of Brig, near the Swiss-Italian border.

It was designed by the Swiss engineer Christian Menn. His father had been a close associate of Robert Maillart (see pp. 120-1), and that master's influence clearly shows in some of Menn's earlier bridge designs. Overlaid on the inspiration from Maillart, however, were the design consequences of the availability of prestressing, as well as the method of progressively cantilevering construction developed by Ulrich Finsterwalder.

Menn's previous major bridge, built at Felsenau in 1974, had a curved roadway, which Menn carried on wide cantilevers from slightly arched, fairly deep and narrow concrete box-girders atop pairs of thin prestressed columns – all combining to create an airy and sinuous appearance.

The road to be carried across the Ganter Valley was also to be curved – in this case doubly so in a shallow S – so Menn designed the bridge so that a straight main span of 174m (571ft) was flanked by oppositely curved 127m (417ft) side-spans. The depth of the valley required one of the piers to be 150m (492ft) high, which was far too much for slender columns such as those at Felsenau, so Menn turned the necessity for massive, hollow, vertical boxes into a virtue by reducing the deck depth as much as possible, thereby emphasizing an opposite contrast to the one at Felsenau.

This shallow deck required support beyond that which could be supplied by its own rigidity, so Menn incorporated cable stays but with a difference. Those supporting the curved side-spans had themselves to follow the curved plan, and he therefore coaxed them into shape by encasing them in thin curved concrete walls on either side of the roadway, matching the effect visually (although strictly speaking, structurally unnecessarily) over the central span. As well as introducing the incidental benefit of corrosion-protection for the cables, this also reduced the stresses on them by fixing them to the walls.

The Ganter Bridge is thus not really a new structural type, although it does incorporate significant technical innovations. Unquestionably, however, its unique profile provides a quite different visual and aesthetic experience from that of any comparable work, complementing its glorious site as one of the great bridges of the 20th century.

Facing page **The grandeur of the Swiss Alps forms a fitting backgound to the main span of the Ganter Bridge, whose profile is rendered unique by its concrete-sheathed cable-stays. On the right of the picture the road curves into the side-span from the valley-side and here, where the furthest pier stands, the road is steep, and steepens even more as it descends towards the Ganter torrent. The nearer pier is nearly twice as high as the further pier, making it virtually as lofty as the main span is wide.**

Left **The taller of the two main piers viewed from the valley floor, and showing the sharp curve of the roadway into the valley wall.**

Right **A side view of the pattern of supporting cables for the main spans of the Ganter Bridge, and of the position of their anchorage inside the concrete sheaths.**

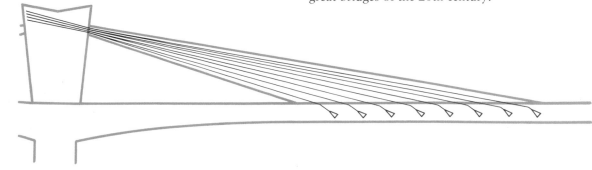

THE HUMBER BRIDGE
ENGLAND

Suspension Spans 2
For earlier spans (1-4), see p. 84.
For later spans (9-11), see p. 162.

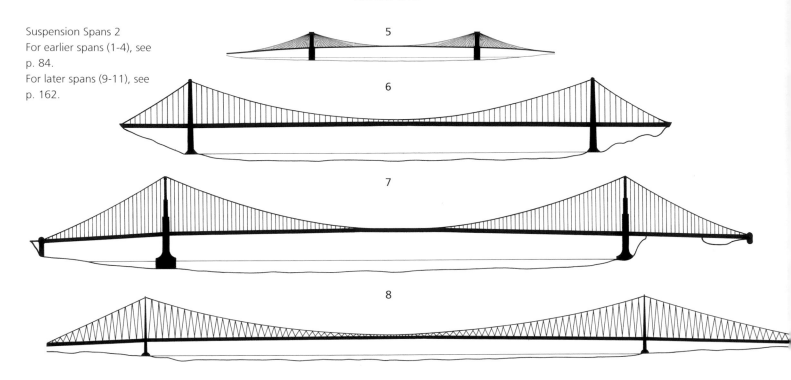

The Severn Bridge conclusively demonstrated the effectiveness of the aerofoil deck design; and its designers used it again on their next major suspension bridge, which literally linked continents, joining Europe and Asia across the Bosporus at Istanbul.

This first Bosporus bridge had a span of 1,074m (3,523ft) suspended from towers both founded on land, with relatively short approach viaducts instead of suspended side-spans. The cables were merely anchored back, at distances of 231m (758ft) and 255m (837ft), beyond each extremity of the main span. The bridge, constructed entirely of steel as were all long-span suspension bridges up to that time, took only three years to complete. It was then the longest span in Europe and the fourth longest in the world (a slightly longer second Bosporus suspension bridge was completed nearby in 1988). With its success it followed the Severn Bridge as the second precedent for the span that was to set a new world record for length – 1,410m (4,624ft) across the Humber Estuary on the north-east coast of England.

Plans for a link across this estuary had come and gone for over a century. Indeed, Freeman Fox had originally proposed a multi-span truss bridge in 1927, but anxiety about the action of the multiple piers on the shifting river-bed eventually led to the adoption of one vast suspension span. Work began in 1972 but, unlike the relatively trouble-free construction of the Bosporus Bridge, it was plagued

Above (top to bottom)
Twentieth-century advances in suspension spans as seen against the Brooklyn Bridge (5). The George Washington Bridge (6) more than doubled the Brooklyn Bridge main span in 1931, and in 1937 the Golden Gate (7) was a further significant extension. The Humber Bridge (8) had the longest main span from 1981 to 1998, but with widely differing side-spans of 280m (919ft) and 530m (1739ft).
Facing Page **As on all modern suspension bridges, the main cables of the Humber Bridge were air-spun, and the supporting strands were inclined in the "zig-zag" form rather than in the longer-established parallel American arrangement.**

from the outset by construction difficulties, severe weather, political controversy and spiralling costs, and it took eight long years to complete. The key problem lay in the founding of the south (Barton) pier. Like both of those at the Bosporus, the north (Hessle) tower had its foundations on land, but even with the huge main span, the south pier still had to be placed over 500m (1,640ft) out into the estuary, where there were 30m (100ft) of shifting sands below the river-bed, and beneath these not bedrock but treacherous clay.

The twin caissons for the Barton tower had to be sunk even further into the clay layer than had been anticipated, and repeated setbacks on this alone delayed the programme by almost two years. For the first time on a long-span suspension bridge, the towers were built of concrete. Again unlike the Bosporus, where the 64m (210ft) deck clearance needed for large ships pushed the tower height up to 165m (541ft), a 30m (100ft) clearance was enough. As a result, the towers could be 10m (33ft) shorter, despite the 30% longer span; and the effect of this, enhanced by the thinness of the steel aerofoil deck, is to make the Humber Bridge look daringly insubstantial and breathtaking even by the standards of other long-span suspension bridges. Both the cable-spinning and the deck erection were beset with further problems and the Humber Bridge was not officially opened until July 1981.

THE PONT DE NORMANDIE, BRITTANY
FRANCE

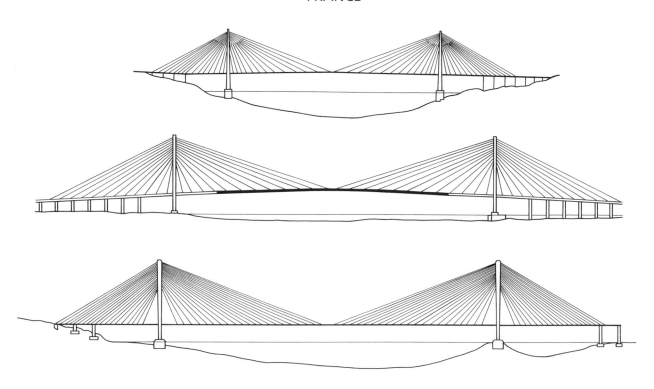

The Humber Bridge spans some 8.6 per cent further than the previous suspension record-holder, the Verrazano Narrows Bridge. This is rather more than the slender 1.4 per cent by which the latter exceeded the Golden Gate, although it is modest against the 20 per cent by which San Francisco's great landmark surpassed the George Washington. It is necessary to take one further step back, however, to the "quantum leap" which the George Washington's designer Othmar Amman took beyond Detroit's Ambassador Bridge in 1931, in order to find a jump in scale for suspension spans comparable with the current "state-of-the-art" in cable-stayed structures – the mighty bridge which now stands in northern France.

The Pont de Normandie, completed across the mouth of the River Seine near Le Havre in 1994, surpasses the Yangpu Bridge by 42 per cent, with a main span of 856m (2,808ft) anchored from inverted Y-shaped towers 214m (702ft) high. The tower shape is similar to that of the Skarnsundet Bridge, though the vertical element is longer and more pronounced, with all the cables fanning out from anchors in this steel, concrete-encased section, rather than from the straddling prestressed-concrete legs beneath. Just as much as with the Norwegian bridge, the shape favours stability – which is crucial for this even more far-reaching and slender structure.

The whole bridge, designed by the French road administration design office SETRA with

Above **Testing the limits of cable-stayed designs. The Skarnsundet Bridge, Norway (top) (pp. 134-135), the previous but one record-holder for length of span in a cable-stayed bridge (at 530m/1,739ft), seems small in comparison with the Pont de Normandie (centre), completed in 1994 (856m/2,808ft), which truly tests the "state-of-the-art" in this technique. Japan's Tatara Bridge (bottom), not due for completion until 1999 (see pp. 160-61), will have only a slightly longer main span – (890m/2,919ft).**

COWIconsult, is over 2km (1¼ miles) long, with extensive approach viaducts on both sides. The southern viaduct has 11 spans; the northern 15. The final 90m (295ft) of the approach spans and the first 116m (380ft) of the main span on both sides, all made of prestressed concrete, were built out from the towers by the balanced cantilever method. The side-spans were connected to the viaducts for maximum rigidity on each side, and then work started on launching out the middle section of the main span, which is 23m (75ft) wide and 3m (10ft) deep. This central 624m (2,047ft) beam is of steel for lightness and strength; and as its two halves grew ever longer and closer to their meeting-point, the formidable problem of keeping the two great ribbons of steel stable in high winds had to be overcome. The first intention was to anchor their ends by cables to points on each side in the river when the wind reached a certain intensity, but this would have been a potential danger to shipping. The method chosen was to include a massive, 50-ton counterweight on each deck and make it capable of moving in any direction to dampen the deck's natural frequency of vibration. It was a step into the unknown, and the designers and builders of this unprecedented structure did not know until they tried whether these "tuned mass dampers" really would control two near half-kilometre lengths of freely cantilevering suspended steel roadway.

Inset above **Prefabrication under way for the concrete units of the approach spans of the Pont de Normandie. A tower crane stands ready between the bases of the land-based south tower.**

Left **The progress made by autumn 1992. The 547m (1,795ft) 11-span southern viaduct, nearest the camera, approaches the tower, at 145m (476ft) still only two-thirds of its full height. The north tower is at approximately the same stage of construction. In the distance the piers for the 737.5m (2419ft) northern viaduct await the installation of the deck elements.**

THE HONSHU-SHIKOKU BRIDGE PROJECT: THE PRESENT
JAPAN

Japan consists of four main islands: Hokkaido and Kyushi lie at opposite ends of the largest, Honshu, which encloses the smallest, Shikoku, in a vast bay, called the Seto Inland Sea, on its southern side. Three undersea tunnels and a suspension bridge link Honshu and Kyushi; the world's longest railway tunnel connects the other end of Honshu with Hokkaido; and during the last quarter of the 20th century what has to be the world's most ambitious civil engineering project is in the process of providing no less than three separate routes between Honshu and Shikoku. The Honshu-Shikoku Bridge Authority was established in 1970 to construct and maintain this final link-up between all the major parts of the country.

The outer two routes, Kobe to Naruto and Onimichi to Imabari, were both begun before the third, the Kojima to Sakaide route, but the latter complex was completed first (by 1988) in an astonishingly sustained design and construction programme. Three large suspension and two cable-stayed bridges, three viaducts and a steel-truss bridge, all double-decked and carrying both road and rail routes throughout the length of the complex, were built across a series of small intermediate islands like giant stepping-stones in just ten years – an achievement made all the more remarkable by the fact that no really long-span bridges had up to that time been built in Japan.

The first under construction were the southernmost main elements, twin suspension bridges named the Kita and Minami Bisan-Seto Bridges, with main spans of 990m (3,248ft) and 1,100m (3,609ft). Both have relatively short side-spans of a uniform 274m

Above and facing page, bottom **The Shimotsui-Seto suspension bridge leads via the Hitsuishijima Viaduct to the twin cable-stayed Hisuishijima and Iwakurojima Bridges and beyond. Amazingly, the Shimotsui-Seto, with its 149m (489ft) steel towers carrying the double-deck truss over 30m (100ft) above water level, is just one small part of the largest civil engineering project in history. See pp.160-1 for the other planned links.**

(899ft), the adjacent pair meeting in a common anchorage, an immense concrete block towering 85m (279ft). The deck had to be almost this high above the water to allow for the passage of the tallest ships and, with both road and rail to be accommodated, there was no question of it being the slender, British-designed, aerodynamic type. Instead, the traffic is carried in a massive steel truss 35m (115ft) wide and 13m (43ft) deep.

On either side of the Kita and Minami Bisan-Seto Bridges there are viaducts, the Bannosu forming the connection with Shikoku, and the 717m (2,352ft) Yoshima Viaduct leading directly to the four steel-truss spans of the Yoshima Bridge. Then, to the north, come perhaps the most striking of all the structures on the Kojima-Sakaide route, the exactly-paired Hitsuishijima and Iwakurojima Bridges, each with cable-stayed main and side spans of 420m (1,378ft) and 185m (607ft). Although a number of longer cable-stayed bridges have been constructed, there are no other such pairs, and none of the type which support a deep, trussed double-deck. Furthermore, only in Japan would there be towers of this shape, flaring outwards at the top in imitation, perhaps, of the helmets which were worn by medieval Japanese warriors.

Northwards beyond the Hitsuishijima Bridge stands the Hitsuishijima Viaduct, the longest of the three viaducts at 1,316m (4,320ft), and then, finally, the last and smallest of the suspension bridges, the Shimotsui-Seto, although even this has a formidable 940m (3,084ft) main span.

Above **Spiralling access ramps lead from the Yoshima Viaduct down to a spacious parking and recreational area on this small island in the Seto Inland Sea. There is room for 1,000 cars, and facilities for their occupants to view from this vantage point both the Seto Inland Sea National Park and the tremendous network of bridges (2 on the map) that traverses it.**

13

INTO THE FUTURE

In 1987 the Japanese organized an international competition for an "Image of the Bridge of the Future". The aim was "to transcend the importance of bridges as transportation means and to define a future image that rediscovers the meaning and effectiveness bridges have for human beings within the space of the world of nature and civilization". Simultaneously, the British ran a parallel competition. In total, well over 700 entries were received.

The key word was "Image". Many entries were drawn as visual metaphors; others depicted "real" structures. The winner of the Japanese competition, the architect Yuri Borzov from the USSR, drew a "Bridge for the Dwellers of Forests and Valleys" –

expanses of "natural" landscape cloaking vaulted subterranean spaces for stores, garages and so on, all separating and filling the distance between Manhattan-like concentrations of glass skyscrapers: a brilliant joke with a serious sub-text, subverting and inverting received notions of what a bridge is.

Many other innovative ideas were entered – an open tube of cables containing a bridge deck; a twisted truss; an arch formed from a tubular space frame with transparent cladding, tied by upwardly-curving flexible cables beneath; a "festooned" suspension bridge with two "levels" of cabling having different sag/span ratios, which it was claimed would give adequate stiffness to multi-span suspension structures; a combination suspension/cable structure with successively smaller triangular elements enclosing a deck, all supported from wide-straddling A-frame towers; a disconcertingly Roebling-like "cablenet" suspension bridge; one indescribable *mélange* of part-arch, cantilevering trusses, cables and towers; a multi-jointed retractable crane-like structure to lift a single person in any direction . . . .

Many of even such "practical" entries could never be built with conventional materials: the stresses and strains generated within these technical gymnastic displays would be far beyond the capacity of the finest present-day, high-strength steel. Nevertheless, new materials are already being developed which their advocates believe will lead the way to the true, unknowable, bridge designs of the future – composites of very high-strength, lightweight, artificial reinforcing fibres embedded in other lightweight solids. The fibres can be glass, near-pure carbon, or aramid (an organic compound); and the matrix through which they run can be of plastics or resins.

The "limiting span" for a bridge of any material or form is that which can support its own weight without collapsing. Maximum theoretical spans for high strength steel have been calculated for several structural types: truss, approx. 500m (1,640ft); arch, approx. 1,500m (4,920ft); cable-stayed, approx. 2,500m (8,200ft); suspension, approx. 5,000m (6,400ft). It has been estimated that the limiting span for a carbon or Kevlar (aramid) suspension span would be 12km (7½ miles) – more than eight times the span of the Humber Bridge.

Obstacles remain before such composite materials start to take the place of steel in really large structures – if they ever do. Cost is a major problem; the need for exhaustive and long-term durability testing is another; as is the fact that such fibre composites, being so much lighter than steel, would bring their own stability difficulties. Also, a bridge eight times longer than the Humber would never be eight times wider or thicker – could such a long, thin, flexible ribbon ever be truly safe? Such problems will surely be addressed, overcome, or circumvented by bridge designers of the 21st century. A future genius may even prove they were not problems at all.

Meanwhile, the possibilities of conventional concrete and steel prestressing are far from exhausted. A 127m (427ft) footbridge was recently completed across the Sacramento River in California with a span to depth ratio of 334:1 – slender enough, *but it has no supports*. Precast concrete deck units were threaded onto prestressing cables which were then tightened, leaving a curious but perfectly safe designed-in sag to the deck, which made it in effect a suspension bridge with the deck and cables as one.

Structural ingenuity is thus very much alive, especially when not inhibited by immediate practicality, as in the 1987 competitions. Nevertheless, neither those visions nor the full fruition of the new materials seems destined yet for reality.

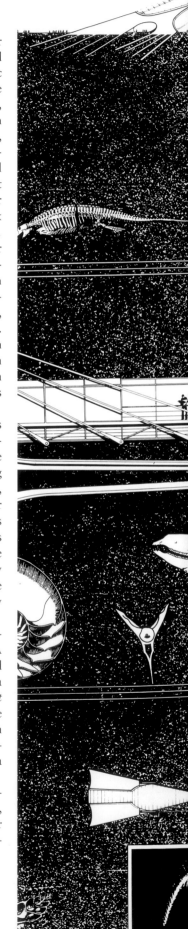

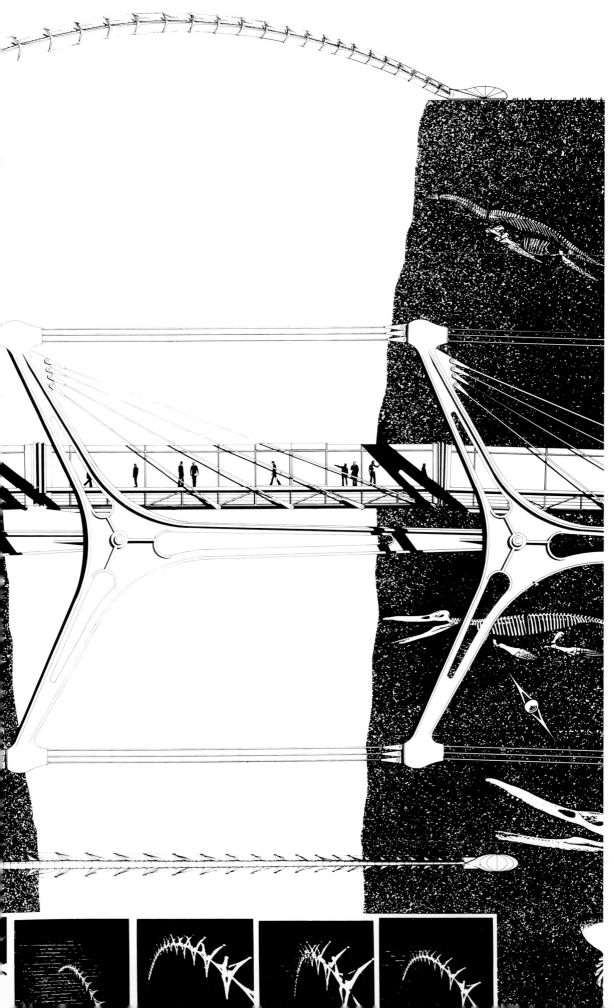

Left **The winner of the British "Image of the Bridge of the Future" competition, designed by David Marks and Julia Barfield with structural advice from Jane Wernick of Ove Arup & Partners. A jointed structural spine of "vertebra" elements, tensioned by cables joining their extended tips, curve in a shallow, flexible arch from a fixed anchorage on one side to a point on the other with only vertical support. A transparent pedestrian travelator is suspended by more cables from the vertebra tips. The concept was explicitly derived from examples of animal spines, and the imagined location for the bridge was the Grand Canyon in Arizona.**

Facing page **This small, unassuming 63m (207ft) span cable-stayed bridge for Aberfeldy Golf Club in Scotland, designed by Maunsell Structural Plastics, is the world's first bridge built entirely in composites – plastic reinforced with glass and cables of Kevlar aramid fibre in a polyethylene coating.**

THE ART OF SANTIAGO CALATRAVA

Since the early 19th century the words "architecture" and "engineering" have implied increasingly separated areas of concern, with the former focusing more and more on issues of aesthetic style, and the latter on the "nuts-and-bolts" of putting structures together.

So can a bridge be "architecture" as well as "engineering"? Some 20th century designers have sought to close, or reclose, the gap. The work of Robert Maillart brilliantly fused style and structure (see pp. 120-1); and the Spanish architect/engineer Santiago Calatrava, a pupil of Christian Menn, has fittingly continued the tradition. Nevertheless, although Calatrava is already widely regarded as the most gifted designer of the late 20th century, the fact that he is not yet 50 earns him a place not so much in the "State of the Art" but in "The Future", as it is most likely that his greatest work still lies ahead of him.

He has already executed many bridge designs, some built, to which "breath-taking" is the unavoidable response. Again and again, Calatrava combines materials and developing forms in organic flights of fancy that extend the limits of the possible in subtle but, once seen, inevitable ways. The bowstring arch as a structural device is familiar enough, but several of his bridges incorporate it in ways that make it seem newly-invented. The Felipe II Bridge in Barcelona has two pairs of steel bowstring arches sweeping obliquely over railway tracks either side of an upward-arching roadway. The outermost of each pair cants inwards sharply to join the other at the apex, creating sidewalks flanked with rows of suspension rods like slender, leaning steel colonnades.

This page **The proposed new steel cable-stayed Medoc Bridge over the River Garonne in Bordeaux is intended to link new developments on the outskirts of the city, and if constructed would be its southernmost bridge.**

The fixed centre bearing in the main drawing is the pivot on which the bridge would swing. While it swings, the span would be stabilized by balance wheels; and when it was being used by traffic, the span would be immobilized.

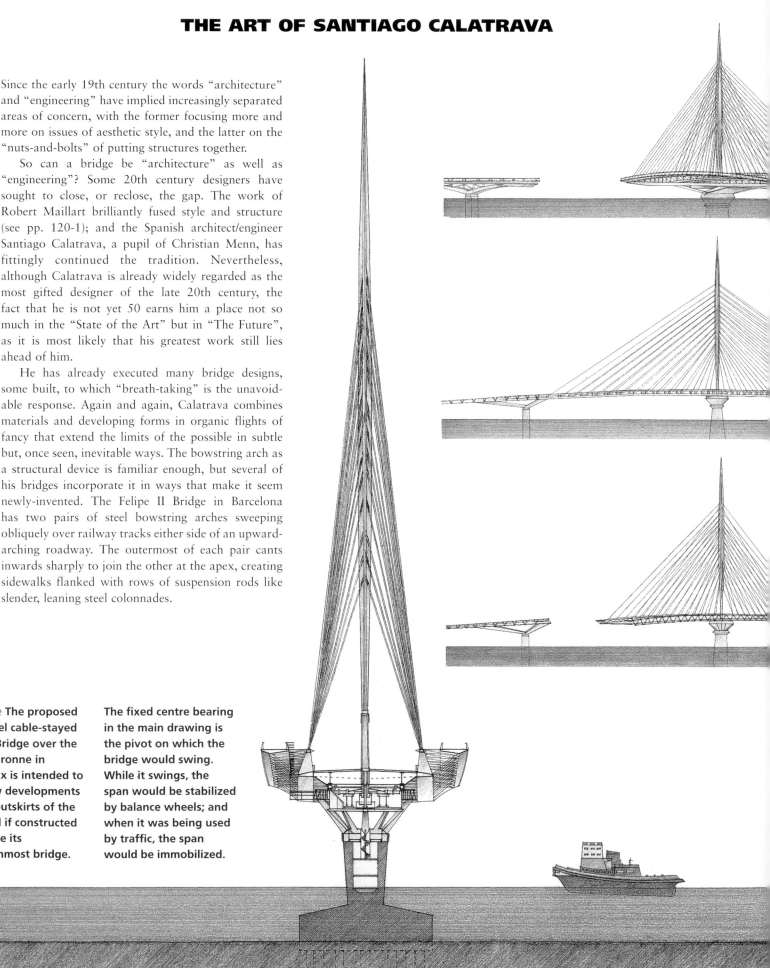

These, and the inner line hanging vertically from the inner arches, do not so much support the roadway as catch it up between them and embrace it.

In Calatrava's unbuilt scheme for the East London River Crossing a single arch would soar across the River with a clear span of 260m (850ft), diving down on each side to piers, then up and out again to meet the six-lane roadway, slicing up and inwards through the roadway's gentle curve to create the central bowstring, like a great steel coat-hanger in the sky. The scheme was dismissed on cost grounds, but at the end of 1992 a campaign was in progress to have it reconsidered. By contrast, the deck suspended from the 195m (640ft) arch of the Merida Bridge in Spain, which is now complete, seems to hug the water surface in its perfectly flat line, at once complementing, challenging, and commenting upon the Roman bridge that crosses the river about 600m (2,000ft) upstream.

All three schemes, like some other arch designs by Calatrava, include cable-staying elements, but his work has also contributed significantly to "classical" cable-stayed designs. The proposed Medoc Bridge at Bordeaux over the River Garonne would be suspended in two equal 120m (400ft) spans from a single central 100m (330ft) mast, about which the entire bridge would pivot to create one of the world's most elegant swing bridges. Whether it will be built remains open at the time of writing.

Calatrava's most celebrated bridge design to date, however, is very much in existence – indeed, it is one of the key images of the Expo '92 at Seville. The Puente del Alamillo suspends its 200m (656ft) span by 13 twin cables from a 142m (466ft) steel tower angled backwards at 58 degrees. Extravagant, expensive, but once seen, unforgettable, the bridge seems to point a finger skywards to the infinite possibilities of the future.

Left The bridge partially opened. The two balanced 120m (394ft) spans would pivot around the 100m (330ft) mast.

Left The bridge in place across the river. Calatrava envisaged it as recalling the silhouette of an old "windjammer" sailing ship.

Left The bridge partially opened in the opposite direction.

Above **Three views showing the inclined pairs of bowstring arches of the 68m (223ft) bridge completed in 1987, which Calatrava designed to link the main streets of Bach de Roda and Felipe II in Barcelona.**

Left The bridge at maximum opening, with the deck lying parallel to the riverbank, allowing shipping to pass on both sides. Calatrava's design included cantilevering the sidewalks upwards and outwards from road level, with gaps through which motorists could enjoy the view.

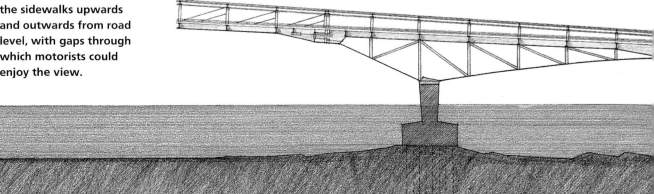

THE GREAT BELT LINK

DENMARK

The Storebaelt (or Great Belt) is the 18km (11½ mile) sound between Denmark's two largest islands, Funen and Zealand. From the west coast of Funen, two bridges built in previous decades already connected it with the mainland of Jutland across the Lillebaelt (or Little Belt). To the east, however, Zealand – which contains the capital, Copenhagen, and half the country's total population – had been accessible only by sea or air. From the mid-1980s, however, preparations were under way for the construction of a fixed Great Belt link, the brainchild of Danish engineer Niels Gimsing, with a length totalling nearly double that of the first Honshu-Shikoku bridge system in Japan (see pp. 152-3).

The tiny island of Sprogø lies roughly half-way across the sound and provides a clear dividing point between the elements of the link. Structures in steel, concrete, and a double-deck steel and concrete composite were all considered for the 6.6km (4 mile) road and rail West Bridge, but after much re-evaluation and redesign, the West Bridge was built as two bridges, each with 51 110.4m (374ft) and 12 81.75m (268ft) concrete box-girder spans. The 25m (82ft) wide road bridge is carried on a shallower box than the 13m (42½ft) rail bridge, and the whole is very gently

arched to allow for 19.75m (63ft) clearance under two navigation spans. The entire structure was prefabricated in 324 huge elements, many weighing up to 6,000 tons, and placed from a giant catamaran crane nicknamed "the Swan". Site work on the West Bridge began in 1990, and was completed in January 1994.

Opening of the rail part of the West Bridge depended, of course, on completion of the remainder of the line, which from Sprogø dives into the twin, mostly bored, East Tunnel for the rest of the link to

Right **Part of the drastically remodelled island of Sprogø. The approach from the road and rail West Bridge will arrive at its far end, where road and rail will part company. The rail line will continue straight ahead into the East Tunnel, while the roadway (which curves into the foreground) will be carried aloft by the western approach viaduct towards the mighty East Bridge.**
Left **Some of the caissons for the East Bridge's anchor blocks being precast.**
Facing Page **A model of the main East Bridge, whose deck is an aerofoil. It consists of a continuous welded steel box-girder suspended by vertical hangers 24m (79ft) from the twin 85cm (33½ in) main cables. These are anchored down to colossal wedge-shaped concrete block caissons (in the foreground of the model) 2.7km (1½ miles apart), after passing over the tapering prestressed-concrete towers – at 254m (833ft) the first to surpass in height those of the Golden Gate. The map shows how the bridge links Funen and Zealand.**

Zealand. After major delays with the East Tunnel, including fire and flood, rail traffic began in 1997. Road traffic, on the other hand, continues onto what makes the whole Storebaelt project a major landmark in bridge-building history. A western viaduct, 1,552m (5,092ft) long, carrying a four-lane dual carriageway plus two emergency lanes, rises gently over 13 spans to meet the world's second largest suspension bridge. This central part of the East Bridge consists of a 1,624m (5,328ft) main span and 535m (1,755ft) side-spans, thus displacing both the Humber and Mackinac Straits from their former records as the world's longest suspended span and overall longest suspended structure and only exceeded by Japan's Akashi-Kaikyo Bridge (see pp. 160-161). The enormous span is necessary because of the huge volume of shipping passing between the Baltic and the North Sea, provided with a navigation clearance of 65m (213ft). From the easternmost anchor block, the road descends over a further 21-span 2,544m (8,346ft) approach viaduct to Zealand.

The principal designers for the Storebaelt link are the Danish engineers COWIconsult, together with several other specialist consultants, including the British Acer Freeman Fox and the German Leonhardt on the East Bridge. Work on this bridge began in October 1991, and it opened in June 1998.

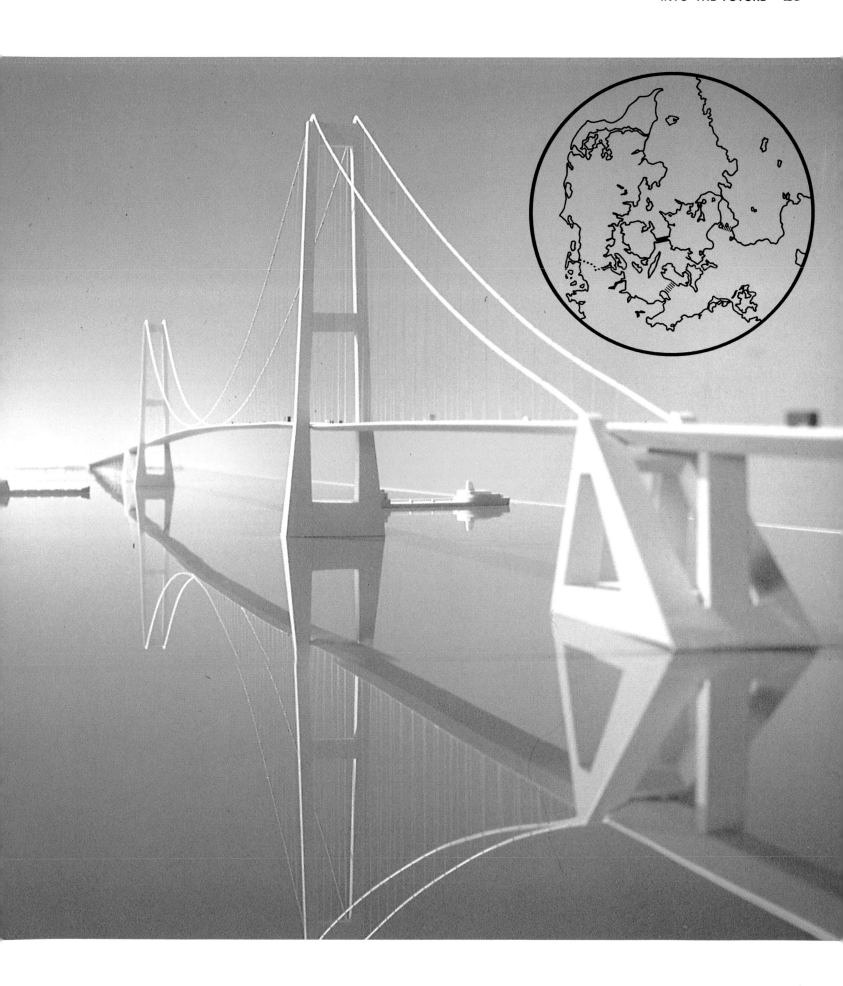

THE HONSHU-SHIKOKU BRIDGE PROJECT: THE FUTURE

JAPAN

Denmark's East Bridge was overtaken in the race to be the world's largest suspension bridge by the second of the three links between the Japanese islands of Honshu and Shikoku. Last to be started but first to be finished was the central Bisan-Seto complex; the link to the west between Onomichi and Imabari which began first with a relatively small cable-stayed bridge in 1968 will be the last to be completed; and between the two in terms of time comes the easternmost Kobe-Naruto Route.

This has two main elements only. The first is the Ohnaruto suspension bridge, which lies to the south, between Naruto on Shikoku and the small intermediate island of Awaji, and which was completed in 1985. Although a substantial double-decked truss structure with main and uniform side-spans of 876m (2,874ft) and 330m (1,083ft), it was dwarfed when its huge counterpart across the 4km (2½ mile) Akashi Strait more than 50km (30 miles) northwards at the other end of Awaji was completed in 1998. This beats all records for span and height. Its total suspended length is 3,911m (12,831ft) and includes a central span of 1,991m (6,532ft). The steel towers are 297m (974ft) high – virtually the same as the Eiffel Tower. While in the West the concept of the slender aerodynamic deck has been applied on almost all long-span suspension bridges built since the mid-1960s, the Japanese have by-and-large opted to retain the older, American-style deep truss for their decks. The Akashi Kaikyo Bridge is no exception and, like many of its fellows, this mighty structure presents a curiously old-fashioned profile to the world. At this location, however, the difficulties for the bridge-builders were immense, and strength and stability for the structure were all-important. The nearby 1995 Kobe earthquake proved its seismic design. The completed towers

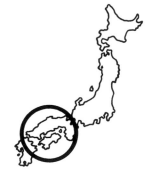

Above **A simulation of the Akashi Kaikyo Bridge. The largest single element in the whole Honshu-Shikoku complex, it spans 4km (2½ miles) of the busiest waterway in Japan. The foundations withstand a current of 4.5m (14.5ft) per second, while the design wind speed for its superstructure is 46m (151ft) per second. Two 1.1m (3ft 7in) cables support the bridge and its six highway lanes.**

were undamaged, though they moved slightly apart, lengthening the span by 1m (3ft 3in). In addition to the fact that Japan is prone to earthquakes, the Strait is congested with shipping, it is subject to strong tidal currents and high winds, and it is also very deep, requiring colossal pier caissons 80m (260ft) across and 70m (230ft) deep.

Although it has no single structure to match the Akashi Kaikyo (or indeed the Danish East Bridge) the third Honshu-Shikoku link, the Onomichi-Imabari Route, will as a whole have exceeded all the others in scale when it is completed. In addition to the final link between the two main islands, a 60km (40 mile), four-lane highway will connect the route with the archipelago of small islands that lie between them, requiring no less than a further ten long-span bridges. Already in use in the north and central parts of the Route are a mixed bag of one suspension, one three-span truss, one steel arch, one steel box-girder, and two cable-stayed bridges; and the whole complex will be crowned with the construction of the world's longest cable-stayed bridge and three consecutive suspension bridges. The Tatara Bridge, with a very similar fan profile to France's Normandy Bridge, will have a main span just 34m (112ft) longer at 890m (2,920ft). The three Kurushima Bridges, totalling 4,105m (13,620ft), will resemble European designs much more closely than previous Honshu Shikoku bridges, since they will have aerodynamically-shaped decks and towers with straight cross-beams, rather than the girder-like X-shaped connectors mostly used else-where on the route. The Kurushima Bridges, which form the southernmost section of the Route, and the Tatara Bridge are scheduled for completion in 1999.

Right **With its long (960m/3150ft), equal side-spans, deep bracing truss, X-configuration towers, and all-steel construction, the Akashi Kaikyo Bridge (simulated here) not only epitomizes the specifically American style of long-span suspension bridges of the first half of the 20th century, but also represents the ultimate in design of this form for the whole century.**

Right **When the Tatara Bridge (simulated here) is completed in 1999 its record-breaking main span will be flanked by slightly unequal side-spans of 270m (886ft) and 320m (1050ft). The Tatara – the central element of the Onomichi-Imabari Route (no. 1 on the map) was originally conceived as a suspension bridge, but after extensive investigation the Japanese concluded that recent progress with the cable-stayed form would make this structure a more economical and time-saving alternative. The Kobe-Naruto Route is marked no.3 on the map. See pp.152-3 for the other Honshu-Shikoku link.**

THE STRAIT OF MESSINA
ITALY

The 3km (2 miles) of sea that divide the island of Sicily from Calabria in southern Italy are relatively modest compared with many of the other stretches that have been bridged. Nevertheless, the great depth of the water, the high winds and the area's susceptibility to earthquakes are all formidable obstacles to bridge builders.

The water depth demands as few piers as possible. Nearly half a century ago, David Steinman envisaged a bridge with one great mile-wide suspension span, which, although much longer than any that had then been built, would still have had to be founded in over 120m (400ft) of water. In 1969 other concepts were submitted in an ideas competition organized by the national corporation Azienda Nazionale Autonoma delle Strade Statali (ANAS), among them a cable-

Facing Page **Stretto di Messina's model of the tremendous structure which may cross the Strait of Messina, Italy, and whose 376m (1,234ft) towers would exceed all but a handful of skyscrapers in height.**

stayed design by Fritz Leonhardt which would have spanned half as far again as any other conceived even today. No scheme was immediately taken up, however, and progress was slow. By 1976 a committee which included the great Italian engineer Riccardo Morandi had laid out a study programme for the crossing; five years later the state-sponsored Stretto di Messina company was formed; and five years after that the company issued a feasibility report on three different ways of crossing the gulf: a suspension bridge, a floating tunnel and a bored tunnel. A mass of evidence, garnered from geologists, seismologists, geotechnicians, experts in sea construction and others, eliminated both tunnel ideas, as well as that of founding any intermediate bridge piers. The most heroic and visionary solution turned out to be the

Suspension Spans 3
For earlier spans 1-4 and 5-7, see pp. 84 and 148.

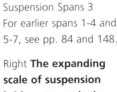

Right **The expanding scale of suspension bridges towards the Third Millennium: the Humber Bridge (8) survived for 17 years as the longest span before Denmark's East Bridge (9) and Japan's Akashi Kaikyo Bridge (10) were completed in 1998. The latter will remain** the longest overall suspended structure by just 250m (850ft) even if the Messina Strait Bridge (11) is built.

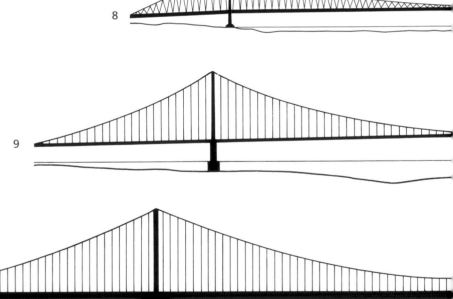

only practical one – the world's longest, single, suspended span, founded on land in Sicily and Calabria.

Meanwhile Morandi, by then in his mid-80s, had played a central role in the development of the single-span solution, although the final design handed over at the end of 1992 adopts a pure suspension configuration for the main span, rather than the suspension/cable hybrid that he presented in 1986. The bridge's colossal length is of course its most remarkable feature – at 3.3km (2 miles) between the towers it will far outstrip the Akashi Kaikyo (although not in total suspended length, as its side-spans are comparatively short).

The next most remarkable feature of the final design is the width of the deck. The bridge is seen not merely as accommodating the existing traffic between Sicily and the mainland, but as the focus and catalyst of a massive urban regeneration of both, perhaps leading to the establishment of a future "City of the Strait". In anticipation of this, the 60.4m (198ft) deck – a vast steel aerofoil nearly twice the width of all other major suspension bridges – is intended to accommodate a pair of rail tracks in the centre plus two service roadways, a triple carriageway with a service lane on each side of these and, cantilevered beyond the line of the bridge hangers, at the far extremity on each side, an emergency road lane positioned some 2m (6½ft) below the level of the remainder of the deck. All this is to be carried some 70m (230ft) above the Strait by multiple hangers at 30m (100ft) intervals, from two pairs of 1.2m (4ft) steel cables stretching over 5km (3 miles) from the anchorage on Sicily to the anchorage in Calabria.

Will the latest "world's longest bridge" be built? The company is convinced that it can be done, and whether economics allow it to go ahead will be known within the lifetime of this book.

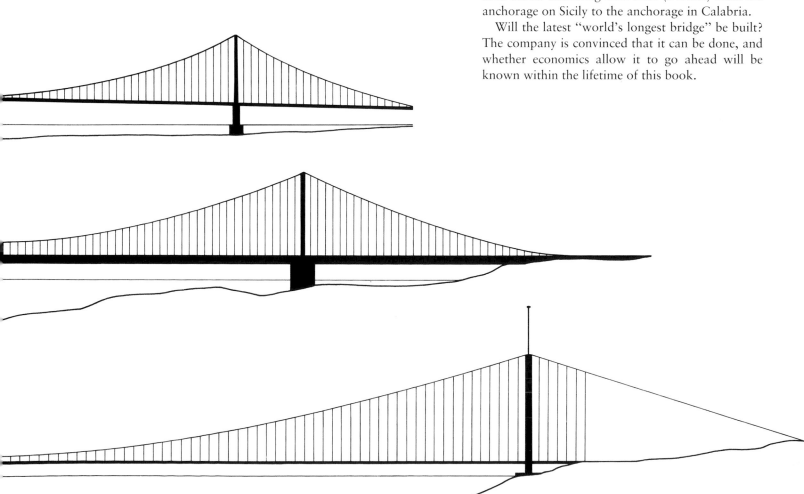

BRIDGES BETWEEN CONTINENTS

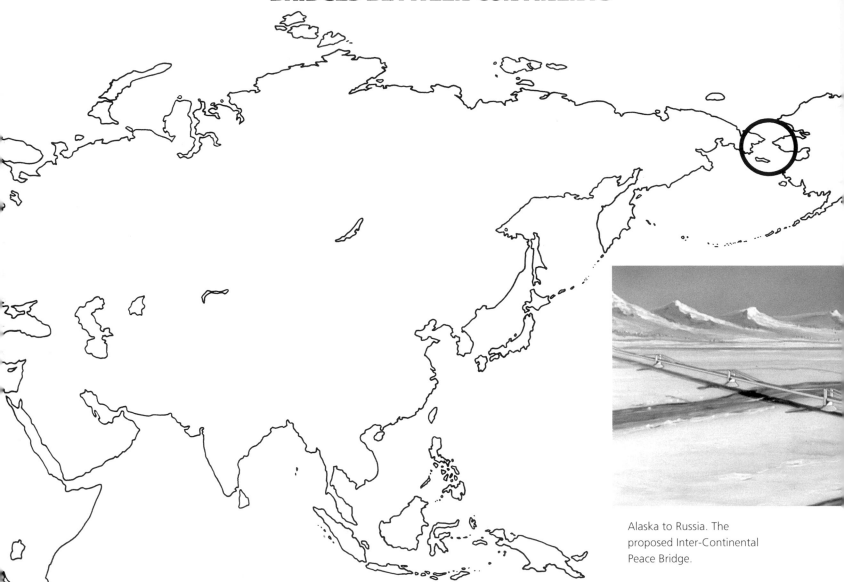

Alaska to Russia. The proposed Inter-Continental Peace Bridge.

The Bosporus, between Asia and Europe, has been bridged several times, but the other possible locations for intercontinental bridges are vastly more challenging and forbidding.

Europe to Africa? The narrowest distance between Spain and Morocco, from Punta Oliveros to Pointe Cires, is only 14km (8½ miles) but the water depth for most of that is far greater than any yet plumbed by bridge piers. A little to the east the distance doubles, but the average depth roughly halves, and here, between Punta Paloma and Punta Malabata, in a study carried out in 1990, COWIconsult located their most practicable site for a bridge link. Given the still-enormous depth of the water, there is an obvious need to minimize the number of deep anchorages for piers, and so a multi-span suspension bridge was the

Above **In the proposed design for the Inter-Continental Peace Bridge a deep prestressed-concrete box section allows each span of the bridge to be supported by a single pair of cables. Twin highway lanes run above the box, with rail tracks on an upper level within it and major service pipes beneath.**

clear choice. The proposal has nine main 2,000m (6,560ft) spans, flanked by 1,000m (3,280ft) side-spans, with 11 200m (656ft) approach spans on the Moroccan side and 26 on the Spanish side.

In 1991 the American Chinese engineer T.Y. Lin proposed an alternative for the shorter route with only two 5,000m (16,400ft) spans and side-spans of 2,500m (8,200ft). For this, the three piers would reach depths of about 80m (260ft), 450m (1,500ft) and 360m (1,200ft). As well as a classical suspension design for the main spans, Lin outlined two hybrid ideas, using part-suspension, either combined with cable-staying or with the cantilever – and it is worth remembering that the potential envisaged by some for fibre composites implies that the Strait here could be crossed by a *single* main span.

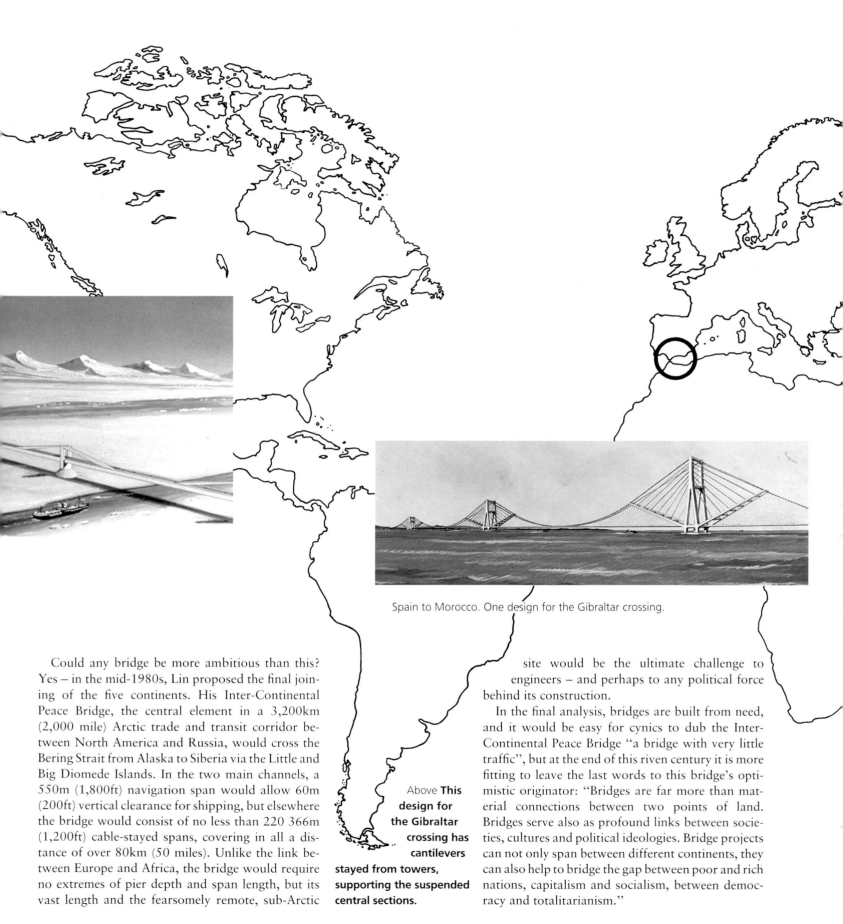

Spain to Morocco. One design for the Gibraltar crossing.

Could any bridge be more ambitious than this? Yes – in the mid-1980s, Lin proposed the final joining of the five continents. His Inter-Continental Peace Bridge, the central element in a 3,200km (2,000 mile) Arctic trade and transit corridor between North America and Russia, would cross the Bering Strait from Alaska to Siberia via the Little and Big Diomede Islands. In the two main channels, a 550m (1,800ft) navigation span would allow 60m (200ft) vertical clearance for shipping, but elsewhere the bridge would consist of no less than 220 366m (1,200ft) cable-stayed spans, covering in all a distance of over 80km (50 miles). Unlike the link between Europe and Africa, the bridge would require no extremes of pier depth and span length, but its vast length and the fearsomely remote, sub-Arctic

Above **This design for the Gibraltar crossing has cantilevers stayed from towers, supporting the suspended central sections.**

site would be the ultimate challenge to engineers – and perhaps to any political force behind its construction.

In the final analysis, bridges are built from need, and it would be easy for cynics to dub the Inter-Continental Peace Bridge "a bridge with very little traffic", but at the end of this riven century it is more fitting to leave the last words to this bridge's optimistic originator: "Bridges are far more than material connections between two points of land. Bridges serve also as profound links between societies, cultures and political ideologies. Bridge projects can not only span between different continents, they can also help to bridge the gap between poor and rich nations, capitalism and socialism, between democracy and totalitarianism."

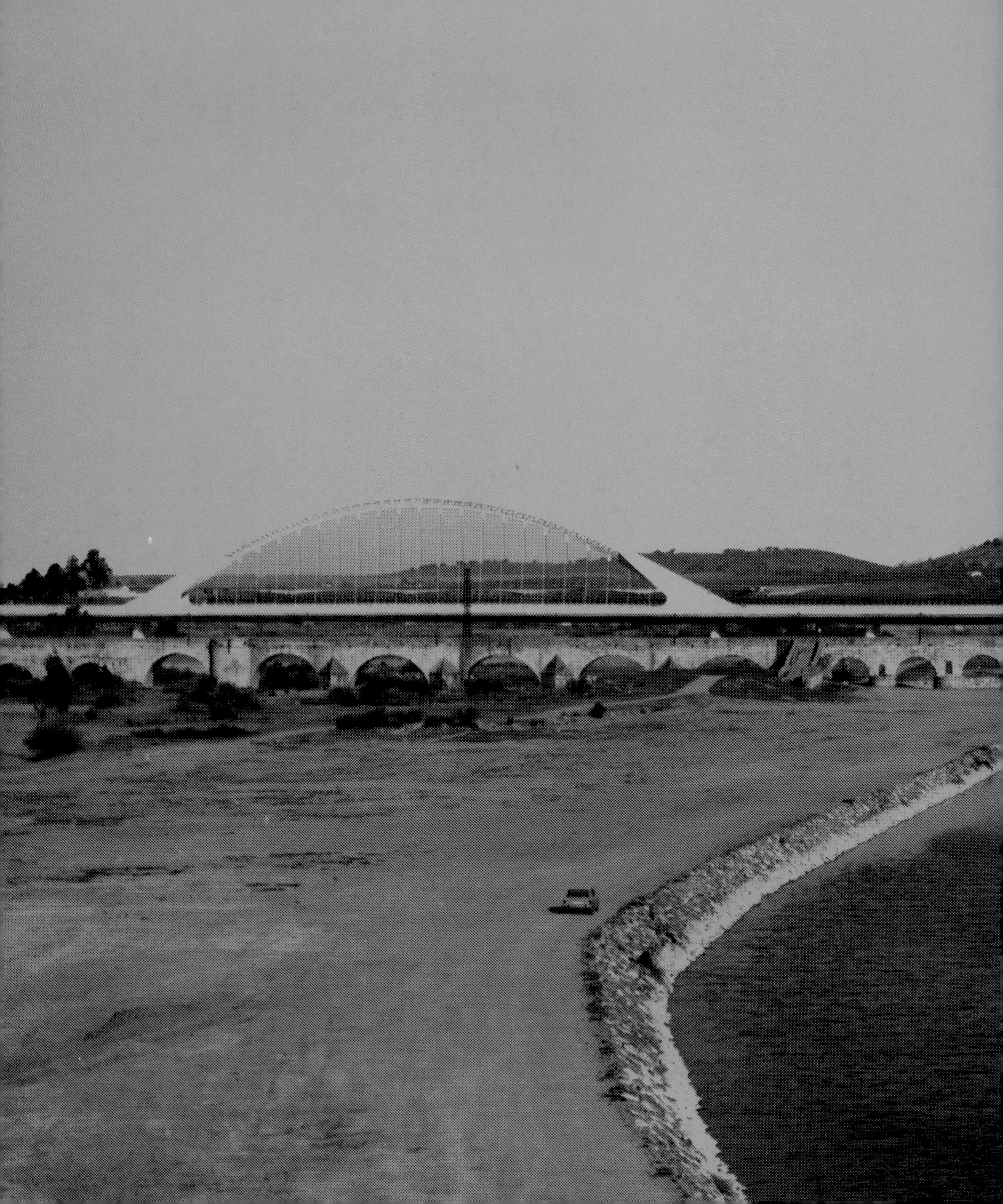

Glossary
Bibliography
Index
Picture Credits

Arches two thousand feet and two thousand years apart: Santiago Calatrava's Merida bridge crosses the Guardiana River, Spain, downstream from the town's 64-arch Roman bridge.

Some of the following terms have more than one common dictionary definition. In such cases, the meanings given here are those most relevant to this book. Key cross-references are given in bold.

Abutment The ground end-support of a bridge, especially to resist the **horizontal thrust** of an arch.

Aerodynamic stability The ability of a bridge deck to withstand wind forces without damage from **torsion**, or **oscillation**: most relevant to **cable-stayed** and **suspension bridges**.

Aerodynamic deck A bridge-deck with a cross-section tapering at each edge to provide **aerodynamic stability**.

Air-spinning A modern method of constructing **suspension bridge cables**, in which wires are continuously unspooled back and forth across a span and bound into strands.

Anchor arm The **side-span**, usually of a **cantilever bridge**, from **abutment** to **pier**, balancing the **cantilever**.

Anchorage A secure fixing, usually in mass reinforced concrete, at the extremity of a **side-span** or **anchor arm**.

Aqueduct A bridge or channel for conveying water, often over long distances.

Aramid An artificial fibre whose exceptionally high **tensile strength** makes it potentially suitable for very long spans.

Arch A curved structural span.

Art Deco A decorative style of the 1920s and 1930s, characterized by streamlined curves and geometrical forms.

Backspan See **side-span**.

Bar chain See **eyebar**.

Bascule A form of moving bridge in which a hinged counterweight at one end of a span falls, causing the deck to rise.

Batter An inclination from the vertical, as in the sloping side of a bridge **pier**.

Beam A rigid, usually horizontal, structural element which may itself form an entire simple bridge.

Bed joint The joint between the radiating elements of an arch.

Bedrock The solid rock layer beneath sand or silt, especially in a river-bed.

Bøllman truss A patent design of overlapping **wrought-iron king-post trusses** plus further diagonal suspension ties.

Bowstring arch An arch whose ends are linked to resist outward thrust.

Box-girder A beam with a hollow square or rectangular section.

Brittle fracture The fracture of **steel** elements at low temperatures.

Burr truss A timber design, combining **king-post** and arch.

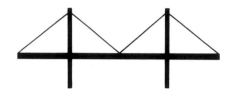

Cable-stayed bridge A bridge whose deck is directly supported from **pylons** by straight cables without vertical suspenders.

Cable The staying or suspending bridge element; in modern **suspension bridges**, the main supporting cable is hung from towers, and formed from steel wire bound in strands.

Caisson A bridge foundation, usually embedded in a riverbed by continuously digging out the material within the bed, so that the caisson sinks.

Camber A slight convexity on the road surface.

Cantilever A horizontal member fixed at one end and free at the other.

Cantilever bridge A bridge with rigid arms extending from both sides of a base, the inner ones usually supporting a central span.

Capital The head of a column in Classical architecture.

Carbon fibre Very high-strength filaments of near-pure carbon, suitable for reinforcement.

Cast iron A brittle alloy with high carbon content: high **compressive strength**, low **tensile strength**.

Catenary The curve into which a uniform

rope or cable falls when suspended from two points, as in a suspension bridge.

Cellular construction In early 20th century American suspension bridges, the method of constructing towers from relatively small welded steel box units.

Cement mortar The mixture of sand, cement, water and lime that binds masonry and brick.

Centering A temporary framework over which arch elements are assembled until they are self-supporting.

Chain The principal supporting element of a now obsolete type of suspension bridge.

Chord The top or bottom horizontal part of a **truss.**

Cladding The outer, usually non-loadbearing, surface of a structure.

Clapper A prehistoric type of stone slab bridge.

Cofferdam A watertight structure allowing underwater foundations to be built in the dry.

Colonnade A series of regularly spaced columns.

Composite construction The use of different materials together in a single structure.

Compressed-air chamber The space at the bottom of a **caisson**, into which air is introduced under pressure to exclude water so that excavation can take place.

Compression The pushing force which tends to shorten a member; opposite of **tension**.

Compression zone The area under **compression** in the upper part of a horizontal beam.

Compressive strength The ability of a material to withstand **compression**.

Concrete A mixture of water, sand, stone, and a binding element which hardens to a rock-like consistency.

Corbelling Successive layers of masonry or brick projecting beyond each other.

Corinthian A Classical architectural style, with leafy decoration at column-heads.

Corne de vache A decorative feature in masonry bridge design, involving shaving the lower curving edge near the **springing** of an arch.

Counterweight See **bascule**.

Creep The slow permanent deformation of material under stress, as in shrinkage of concrete.

Creeper crane The cranes used for building a **steel cantilever bridge**, moving slowly along the upper **chord**.

Crown The highest part of an **arch**.

Cutout The non-structural material removed from a **spandrel**, as in Maillart's bridges.

Cutwater The end of a pier-base, pointed to cleave the water.

Dead load A structure's own weight.

Deflection theory An early 20th-century theory that very long suspension bridges would remain stable without deep stiffening trusses through a balance between flexibility and self-weight.

Doric A Classical architectural style, with no decoration at column-heads.

Dovetail A splayed piece of timber (or iron or stone) fitting tightly into a similarly shaped cutout.

Drawbridge See **bascule**.

Dressing The cutting of stone units to the required shape.

Dry-stone Masonry laid without mortar.

Elliptical arch An arch with a curve that becomes tighter towards the **crown**.

Entablature In Classical architecture, the element that rests upon the **capitals** of the columns.

Environmental load The external forces on a structure, such as wind and water.

Extrados The outer surface of the curve of an arch.

Eyebar The unit from which the **chain** of early **suspension bridges** was constructed, with a flattened ring at each end for linkage.

Falsework Temporary scaffolding during construction.

Fan configuration A **cable-stayed bridge** design in which the cables fan outwards as if from the handle of a fan.

Fender A protective enclosure round a pier structure.

Fill The material, usually rubble or earth, used to fill the space behind the outer surface of a masonry bridge structure.

Fin-back bridge A very modern bridge type in which a vertical solid plane of **prestressed concrete** supports the spans above the deck.

Fink truss A patent design of overlapping wrought-iron **king-post trusses** with additional diagonal bracing.

Flange The flat top and bottom plates of a **box-girder**.

Formwork Temporary boarding to hold concrete in shape while it hardens.

Galvanizing The coating of metal with zinc for waterproofing.

Girder A large **beam**, usually steel or concrete.

Glass fibre A reinforcing material with high **tensile strength**.

Gradient of stress The theoretically uniform change from purely **compressive** forces along the top of a **beam** to purely **tensile** along the bottom.

Granite A hard crystalline rock, suitable for masonry bridges.

Hanger See **suspender**.

Harp configuration A **cable-stayed bridge** design in which cables radiate at a uniform distance from each other throughout their length.

Haunch The part of an arch between the **springing** and the **crown**.

Horizontal thrust The tendency of an **arch** to push outwards.

Howe truss A patent design with vertical iron **tension** rods.

I-beam A **beam** or **girder** with an I-shaped cross-section.

Intrados The inner surface of an **arch** ring.

Ionic A Classical architectural style with scroll decoration at the column-heads.

Jack-knife bridge A form of moving bridge with a deck that hinges upwards at the centre.

Keystone The **voussoir** at the **crown** of an **arch**.

King-post truss A truss consisting of a vertical post, connected to a horizontal beam by inclined tie-beams.

Laminated timber Layers of timber clamped or glued face-to-face.

Lift bridge See **bascule**.

Lime mortar A non-waterproof binding material for masonry, consisting of lime, water and sand.

Limiting span The maximum span possible for each particular type of bridge.

Live load The weight of traffic passing over a bridge.

Long truss A patent timber design based on overlapping **king-post trusses**.

Mortar See **lime mortar** and **cement mortar**.

Mortice A slot in a member, into which a projecting **tenon** is fixed to form a joint.

Navigation span The part of a bridge with maximum clearance for shipping.

Ogival arch A pointed arch

Orthotropic deck A bridge deck which is stiffer in the direction of the span than it is laterally.

Oscillation The movement, usually vertical, of a suspended bridge deck in the wind.

Pier The support between two bridge spans, usually arches.

Pinned arch An arch with **hinges** at the **abutments** and sometimes also at the **crown**.

Plate girder A flat bridge deck with a shallow rectangular section.

Pointed arch An arch with an angle at its **crown**.

Pneumatic caisson A **caisson** with a **compressed-air chamber**.

Pontoon bridge A bridge formed from floating units, sometimes boats, tied together in a series.

Portal A frame with side uprights connected by a horizontal member at the top.

Post-tensioning The method of making **prestressed concrete** with steel strands tightened after the concrete has hardened.

Pozzolana The volcanic dust first found at Pozzuoli, with which the Romans made waterproof concrete.

Pratt truss A patent truss design with iron diagonals in tension.

Pre-tensioning The method of making **prestressed concrete** with steel strands under tension as the concrete sets.

Prefabrication The manufacture of structural units in an off-site factory.

Prestressed concrete A modern type of **concrete** with stretched steel strands embedded in it to impart additional **tensile strength**.

Pylon The vertical structural element from which stays radiate in a cable-stayed bridge.

Reinforced concrete Concrete with steel bars or mesh embedded in it for increased **tensile strength**.

Ripple The undulating motion of a suspended deck caused by wind.

Scour The destructive effect on submerged **piers** from fast-flowing water.

Segmental arch An **arch** formed from a segment of a circle.

Semi-circular arch An **arch** forming a complete half-circle.

Semi-fan configuration A style of **cable-stayed bridge** midway between the **fan** and **harp**.

Shear The force acting across any beam or structural unit.

Side-span The outer suspended section of a **suspension bridge** from the tower to the anchorage, balancing the central suspended span.

Side-sway The movement of a suspended bridge deck from side to side in wind.

Soffit The under-surface of any piece of structure.

Spandrel The area of an **arch** bridge above the **extrados** and below deck level.

Springing The point where the end of an **arch** meets the **abutment**.

Starling The usually boat-shaped foundation for a masonry **pier**.

Steel An alloy of iron with more carbon than **wrought** iron but less than **cast iron**, combining the **tensile strength** of the former with the **compressive strength** of the latter.

Stiffening truss A **truss** usually beneath the entire deck of a **suspension bridge**.

Strand A unit within a **suspension bridge cable**, itself formed from many individual wires.

Striking The action of removing **formwork**, particularly **centering**, from beneath a completed arch.

Suspender The vertical or zig-zag element on **suspension bridge**s that links a cable with a deck.

Suspension bridge A bridge with its deck supported from above by large cables or chains hanging from **towers**.

Swing bridge A type of moving bridge in which the deck pivots sideways.

T-beam A **beam** or **girder** with a T-shaped cross-section.

Tenon A projecting piece of a member that fits into a **mortise** cut in another to form a joint.

Tensile strength The ability of a material to withstand **tension**.

Tension The pulling force that tends to lengthen a member.

Tied arch See **bowstring arch**.

Torsion The strain produced by twisting.

Tower The vertical element in **suspension bridges** from which **cables** are hung.

Town truss A patent **truss** design forming a wooden lattice.

Transporter bridge A type of moving bridge in which a travelling gondola is suspended from an overhead frame.

Trapezoid A four-sided figure with one pair of parallel sides.

Travertine A pale form of limestone.

Truss A frame of members in **tension** and **compression**.

Tuned mass damper A **counterweight** to subdue a bridge deck's tendency to vibrate.

Voussoir The wedge-shaped units, usually stone, from which an **arch** is formed.

Web The side-plates of a **box-girder**.

Whipple truss Several patent designs by Squire Whipple: the most characteristic was a **bowstring**, with a curved **cast-iron** upper **chord** and lower members of **wrought iron**.

Wire cable See **cable**.

Wrought iron Soft and malleable alloy with very low carbon content; low **compressive strength**, high **tensile strength**.

Zig-zag bridge Traditional Chinese bridge type, with deck elements at right angles to each other.

Zig-zag suspension The arrangement of suspension bridge cables first introduced on the Severn Bridge, as differing from vertical suspenders.

BIBLIOGRAPHY

BECKETT, Derrick. **Bridges**. London, Paul Hamlyn, 1969.

BERRIDGE, P.S.A. **The Girder Bridge After Brunel and Others**. London, Robert Maxwell, 1969.

BILL, Max. **Robert Maillart: Bridges and Constructions**. London, Pall Mall Press, 1969.

BILLINGTON, David P. **The Tower and the Bridge: The New Art of Structural Engineering**. New York, Basic Books, 1983.

BOYNTON, R.M. and RIGGS, L.W. Tagus River Bridge. **Civil Engineering**, Vol.36, No.1, pp.34-45, February 1966.

BRACEGIRDLE, Brian. **The Archaeology of the Industrial Revolution**. London, Heinemann, 1973.

BREES, S.C. **Science Pratique des Chemins de Fer**. Bruxelles, Société Belge de Librairie, 1841.

The Bridge Spanning Lake Maracaibo In Venezuela. Wiesbaden, Bauverlag GmbH, c.1964.

CALATRAVA, Santiago. **Dynamic Equilibrium: Recent Projects**. Switzerland, Verlag für Architektur, 1991.

CHATTERJEE, Sukhen. **The Design of Modern Steel Bridges**. Oxford, BSP Professional Books, 1991.

CHATTERJEE, Sukhen, *et al*. Strengthening and Refurbishment of the Severn Crossing. Parts 1-5. **Proceedings of the Institution of Civil Engineers: Structures and Buildings**, February 1992, pp.1-60.

CHRIMES, Michael. **Civil Engineering 1839-1889: A Photographic History**. Stroud, Alan Sutton, 1991.

COLLINS, Dr. A.R. **Structural Engineering – Two Centuries of British Achievement**. The Institution of Structural Engineers Anniversary Publication. Chislehurst, Tarot Print, 1983.

COSSONS, Neil and SOWDEN, Harry. **Ironbridge: Landscape of Industry**. London, Cassell, 1977.

COWAN, Henry. **The Master Builders: A History of Structural and Environmental Design from Ancient Egypt to the Nineteenth Century**. New York, John Wiley. 1977.

DE MARÉ, Eric. **Bridges of Britain**. Revised edition, Batsford, 1975.

DEMPSEY, G. Drysdale. **Tubular and Other Iron Girder Bridges, Particularly Describing the Britannia and Conway Tubular**

Bridges. London, Virtue Brothers, 1862.

Failure Of Tacoma Bridge. Parts 1-4. **The Engineer**, 29 August, 1941 and 5, 12, 19 September, 1941.

Forth Road Bridge Superstructure. A.C.D. Bridge Co., 1965.

FREDRICH, Augustine, ed. **Sons of Martha: Civil Engineering Readings in Modern Literature**. New York, American Society of Civil Engineers, 1989.

GIES, Joseph. **Bridges and Men**. London, Cassell, 1964.

GIMPEL, Jean. **The Medieval Machine: The Industrial Revolution of the Middle Ages**. 2nd edition, Aldershot, Wildwood House, 1988.

GIMSING, Niels. **Cable-Supported Bridges: Concept and Design**. New York, John Wiley, 1983.

HERODOTUS, trans. A. de Sélincourt. **The Histories**. Harmondsworth, Penguin Classics, 1971.

HEYMAN, Jacques. **The Masonry Arch**. Chichester, Ellis Horwood, 1982.

HOPKINS, H.J. **A Span of Bridges**. Newton Abbot, David & Charles, 1970.

HOWIE, Will and CHRIMES, Mike, eds. **Thames Tunnel to Channel Tunnel: 150 Years of Civil Engineering**. Selected papers from the **Journal of the Institution of Civil Engineers**. London, Thomas Telford, 1987.

HUSBAND, H.C. and HUSBAND, R.W. Reconstruction of the Britannia Bridge, **Proceedings of the Institution of Civil Engineers**, Part 1, Vol.58, February 1975, pp.25-66.

INSTITUTION OF CIVIL ENGINEERS. Forth Road Bridge. Papers from **Proceedings of the Institution of Civil Engineers**, by J.K.Anderson, et al. ICE, 1967.

ITO, Manabu, et al. **Cable-Stayed Bridges: Recent Developments and their Future**. Proceedings of the Seminar at Yokohama, Japan, 10-11 December 1991. Amsterdam, Elsevier, 1991.

KEMP, Emory L. Ellet's Contribution to the Development of Suspension Bridges, **Engineering Issues: American Society of Civil Engineers Journal of Professional Activities**, Vol.99, No.PP3, pp.331-351, July 1973.

KIRBY, Richard S. et al. **Engineering in History**. New York, Mcgraw-Hill, 1956.

LEONHARDT, Fritz. **Bridges: Aesthetics and Design**. London, Architectural Press, 1982.

METCALF, Leon. **Bridges and Bridge Building**. London, Blandford, 1970.

MUSGROVE, John, ed. **Sir Banister Fletcher's "A History of Architecture"**. 19th edition, London, Butterworths, 1987.

NELSON, Gillian. **Highland Bridges**. Aberdeen University Press, 1990.

O'CONNOR, Colin. **Design of Bridge Superstructures**. New York, Wiley-Interscience, 1971.

PACEY, Arnold. **Technology in World Civilization**. Basil Blackwell, 1990.

PALLADIO. **The Four Books of Architecture**. New York, Dover, 1965. Reprint of Isaac Ware's translation of 1738.

PANNELL, J.P.M. **An Illustrated History of Civil Engineering**. Thames & Hudson, 1964.

PARSONS, William Barclay. **Engineers and Engineering in the Renaissance**. Reissued, Cambridge, Massachusetts, The MIT Press, 1968, 1976.

PENFOLD, Alastair, ed. **Thomas Telford: Engineer**. Proceedings of a seminar held at Ironbridge, April 1979. London, Thomas Telford, 1980.

PETERS, Tom F. **Transitions in Engineering. Guillaume Henri Dufour and the Early 19th Century Cable Suspension Bridges**. Basel, Boston, Birkhäuser Verlag, 1987.

PODOLNY, Walter, and SCALZI, John B. **Construction and Design of Cable-Stayed Bridges**. 2nd edition. New York, John Wiley, 1986.

PREBBLE, John. **The High Girders**. London, Secker & Warburg, 1966.

PUGSLEY, Sir Alfred, ed. **The Works of Isambard Kingdom Brunel: An Engineering Appreciation**. London, Institution of Civil Engineers/University of Bristol, 1976.

RAMBØLL, B.J. **The Criss-Cross Web**. Copenhagen, NytNordisk Forlag, 1972.

Report of a Royal Commission into the Failure of West Gate Bridge. Victoria, Australia, C.H. Rixon, Government Printer, 1971.

ROLT, L.T.C. **Great Engineers**. London, G. Bell, 1962.
Isambard Kingdom Brunel. London, Longman, 1957.
Thomas Telford. London, Longman, 1958.
Victorian Engineering. Harmondsworth, The Penguin Press, 1970.

RUDDOCK, Ted. **Arch Bridges and their Builders 1735-1835**. Cambridge University Press, 1979.

SANDSTRÖM, Gösta E. **Man the Builder**. New York, McGraw-Hill, 1970.

SEALEY, Antony. **Bridges and Aqueducts**. London, Hugh Evelyn, 1976.
The Severn Bridge Superstructure. Associated Bridge Builders Ltd., 1966.

SHIPWAY, J.S. Tay Rail Bridge Centenary – Some Notes on its Construction 1882-87. **Proceedings of the Institution of Civil Engineers**, Part 1, Vol.86, pp.1089-1109, December 1989.

The Forth Railway Bridge Centenary 1890-1990: Some Notes on its Design. **Proceedings of the Institution of Civil Engineers**, Part 1, Vol.88, pp.1079-1107, December 1990.

SHIRLEY-SMITH, H. **The World's Great Bridges**. 2nd edition. London, Phoenix House, 1964.

SMILES, Samuel. **Lives of the Engineers**. 5 vols. Revised edition. London, John Murray, 1874.

STEINMAN, David B. and WATSON, Sara R. **Bridges and Their Builders**. Revised Edition. New York, Dover, 1957.

STRAUB, Hans. **A History of Civil Engineering**. London, Leonard Hill, 1960.

TAYLOR, P.R. and TORREJON, J.E. Annacis Bridge. **Concrete International**, Vol.9, No.7, pp.13-22, July 1987.

TROITSKY, M.S. **Cable-Stayed Bridges: Theory and Design**. 2nd edition. Oxford, BSP Professional Books, 1988.

Orthotropic Bridges: Theory and Design. 2nd edition. Cleveland, The James F. Lincoln Arc Welding Foundation, 1987.

WESTHOFEN, Wilhelm. The Forth Bridge. **Engineering Magazine**, 1890. Centenary Reprint, Edinburgh, Moubray, 1989.